I THINK, THEREFORE I LAUGH

I THINK, THEREFORE I LAUGH

The Flip Side of Philosophy

JOHN ALLEN PAULOS

COLUMBIA UNIVERSITY PRESS
NEW YORK

Columbia University Press
Publishers Since 1893
New York   Chichester, West Sussex
Copyright © 2000 John Allen Paulos
All rights reserved

Library of Congress Cataloging-in-Publication Data
Paulos, John Allen.
        I think, therefore I laugh : the flip side of philosophy / John Allen
Paulos. — [2nd ed.]
            p. cm.
        Includes bibliographical references (p.167) and index.
        ISBN 0–231–11914–3. — ISBN 0–231–11915–1 (pbk.)
        1. Logic. 2. Philosophy Humor. I. Title.
        BC71.P38 2000
        190.'207—dc21
99–34799

Casebound editions of Columbia University Press books are printed on perma-
nent and durable acid-free paper.
Printed in the United States of America
p 10 9 8 7 6 5 4 3 2 1

For my wife, Sheila

CONTENTS

PREFACE TO THE
SECOND EDITION

I Think, Therefore I Laugh is the second of my six books, and it has been out of print for a while. When Columbia University Press asked me if I would be interested in reissuing it and writing a new preface, I immediately agreed. With an author's myopic vanity, perhaps, I have always liked this little book, inspired, as it was, by Wittgenstein's quip that a book on philosophy might consist entirely of jokes. Since the book went out of print rather quickly, I've used it as a small quarry, and readers of my subsequent books may recognize bits and pieces of it in them. Moreover, many of the book's concerns are similar to those of my later books: misunderstandings of mathematics and science and of the relation between them, pseudoscience and its appeal, the uses and misuses of probability and statistics, humor and "higher-order" endeavors, the interplay between narrative and numbers.

Although my Ph.D. is in mathematics, specifically mathematical logic, I've always had an interest in analytic philosophy and its puzzles. It seemed to me when I wrote *I Think, Therefore I Laugh*, and it still seems, that the border between such philosophical abstractions and the concerns of everyday life is well worth exploring. The payoff to this exploration is of a largely intellectual sort. Recall one definition of a philosopher: he is the one who attends a conference on crime sentencing guidelines and delivers a paper on the meaning of "time" and the logical dilemma faced by imprisoned accomplices. Since social, economic, and topical issues are not the focus of this book (as they have been in a couple of my later works), there is no compelling temptation to update it. Aside from eliminating a number of infelicities and a few minor mistakes, I have not changed anything.

If I were to do the book over, I would choose a slightly different set of philosophical problems and a different set of jokes and parables and would develop them at a more leisurely pace. The presentation here is a bit relentless—something, something else, and then some other thing. Nevertheless, I reiterate and stand by the book's guiding insight: conceptual humor and analytic philosophy resonate at a very deep level. Did you hear what George Carlin and Groucho Marx said to Robert Nozick and Bertrand Russell? . . .

I THINK, THEREFORE I LAUGH

chapter one

TWO UNLIKELY PAIRS OF MEN

Introduction

Ludwig Wittgenstein, the Austrian philosopher, once remarked that "a serious and good philosophical work could be written that would consist entirely of jokes" (Wittgenstein). If one understands the relevant philosophical point, one gets the joke. This has always seemed to me to be a wise remark, and this book is written in part to exemplify it. The book will contain a number of jokes as well as stories, parables, puzzles, and anecdotes, all of which in one way or another will relate to various philosophical problems. These stories and anecdotes will be linked by some (minimal) exposition and will be loosely integrated by topic. I hope they convey something of the flavor and substance of modern philosophy and dispel the feeling among some that philosophy is some sort of guide to life, a branch of theology or mathematics, or merely a matter of being stoical in the face of adversity.

One obvious criticism of an endeavor such as this is that for the philosophical points to be comprehensible, the jokes, examples, and metaphors relating to them must be placed in a relevant context and must be made part of a tightly reasoned argument. This is often true, of course, but for most of them the context and argument are at least partly implicit in the stories themselves. Consider, for example, the story of monkeys randomly typing on a typewriter and *King Lear* resulting. Even with no context or argument, the isolated story is thought-provoking, no matter that the "wrong" thoughts are often provoked. Similar remarks can be made about other classic stories—the sound of a tree falling in an uninhabited forest, Laplace's deterministic image of the universe as something like a giant and inexorable clock, or Plato's metaphor of the cave and the vague reflections of reality it allows. Often what one retains from a philosophical discussion are just such stories, vivid metaphors, examples, and counterexamples. The same thing holds for philosophical jokes.

Finally, even without much supporting context or argument, these stories and jokes are such that any fuller discussion or theory must accommodate and account for them. They provide part of the raw material that any reasonable philosophical theory must make sense of and thus should be part of the intellectual gear of all curious human beings.

Wittgenstein and Carroll

Let me consider a couple of unlikely pairs of men: the first, Wittgenstein and Lewis Carroll; the second, Bertrand Russell and Groucho Marx. The first pair I also compared in my previous book, *Mathematics and Humor*, from which this subsection is taken. However, in this book, among much else, I expand a bit on the comparison, as well as on a few other points made in *Mathematics and Humor*.

George Pitcher in "Wittgenstein, Nonsense, and Lewis Carroll" has written of some very striking similarities between the philosophical writings of Wittgenstein and the work of Carroll (Charles Lutwidge Dodson). Both men were concerned with nonsense, logical confusion, and language puzzles—although, as Pitcher notes, Wittgenstein was tortured by these things, whereas Carroll was, or at least appeared to be, delighted by them. (The relation between the two men is similar in this latter respect to that between Soren Kierkegaard and Woody Allen: same concerns, different approaches.) Pitcher cites many passages in *Alice's Adventures in Wonderland* and *Through the Looking Glass* as illustrating the type of joke Wittgenstein probably had in mind when he made the comment on philosophical jokes mentioned earlier.

The following excerpts are representative of the many in Lewis Carroll that concern topics that Wittgenstein also considered in his writings:

1. She [Alice] ate a little bit, and said anxiously to herself, "Which way? Which way?" holding her hand on the top of her head to feel which way it was growing, and she was quite surprised to find that she remained the same size. (*Alice in Wonderland*)

2. "That is not said right," said the Caterpillar. "Not *quite* right, I'm afraid," said Alice timidly; "some of the words have got altered."

"It is wrong from beginning to end," said the Caterpillar decidedly, and there was silence for some minutes. (*Alice in Wonderland*)

3. "Then you should say what you mean," the March Hare went on.

"I do," Alice hastily replied; "at least—at least I mean what I say—that's the same thing, you know."

"Not the same thing a bit!" said the Hatter. "Why, you might just as well say that 'I see what I eat' is the same thing as 'I eat what I see'!" (*Alice in Wonderland*)

4. "Would you—be good enough," Alice panted out, after running a little further, "to stop a minute just to get one's breath again?"

"I'm *good* enough," the King said, "only I'm not strong enough. You see, a minute goes by so fearfully quick. You might as well try to stop a Bandersnatch!" (*Through the Looking Glass*)

5. "It's very good jam," said the Queen.

"Well, I don't want any to-*day*, at any rate."

"You couldn't have it if you *did* want it," the Queen said. "The rule is jam to-morrow and jam yesterday—but never jam to-day."

"It *must* come sometimes to 'jam to-day,'" Alice object-ed.

"No, it can't," said the Queen. "It's jam every *other* day; today isn't any *other* day, you know."

"I don't understand you," said Alice. "It's dreadfully confusing." (*Through the Looking Glass*)

What do these examples have in common? They all betray some confusion about the logic of certain notions. One does not lay one's hand on top of one's head to see if one is growing taller or shorter (unless only one's neck is growing). One cannot recite a poem incorrectly "from beginning to end," since then one cannot be said to be even reciting that poem. (Wittgenstein was very concerned with criteria for establishing identity and similarity.) In the third quotation the Mad Hatter is presupposing the total inde-pendence of meaning and saying—an assumption that, Wittgenstein shows, leads to much misunderstanding. The next passage confuses the grammar of the word "minute" with that of a word like "train"; and the last illustrates that the word "to-day," despite some similarities, does not function as a date. Both these latter points were also dis-cussed by Wittgenstein.

Wittgenstein explains that "when words in our ordinary language have prima facie analogous grammars we are inclined to try to interpret them analogously; i.e., we try to make the analogy hold throughout." In this way we "misunderstand . . . the grammar of our expressions and, like the fly in the fly bottle, sometimes need to be shown our way clear" (Wittgenstein). As I have mentioned, these linguistic misunderstandings can be sources of delight or of torture, depending on one's personality, mood, or intentions. Wittgenstein, for example, was tormented by the fact that a person does not talk about having a pain in his shoe, even though he may have a pain in his foot and his foot is in his shoe. Carroll, had he thought of it, probably would have written of shoes so full of pain that they had to be hospitalized.

Groucho Meets Russell

Just as Wittgenstein and Lewis Carroll shared some of the same preoccupations with language and nonsense, so Bertrand Russell and Groucho Marx were both concerned with the notion of self-reference. Furthermore, Russell's theoretical skepticism contrasts with Groucho's streetwise brand as do Russell's aristocratic anarchist tendencies with Groucho's more visceral anarchist feelings. I try to illustrate these points in the following dialogue between the two. Some of the topics mentioned in the dialogue will be discussed more fully in later chapters.

Groucho Marx and Bertrand Russell: What would the great comedian and the famous mathematician-philosopher, both in their own ways fascinated by the enigmas of self-reference, have said to each other had they met? Assume for the sake of absurdity that they are stuck together on the thirteenth metalevel of a building deep in the heart of Madhattan.

GROUCHO: This certainly is an arresting development. How are your sillygisms going to get us out of this predicament, Lord Russell? (*Under his breath:* Speaking to a Lord up here gives me the shakes. I think I'm in for some higher education.)

RUSSELL: There appears to be some problem with the electrical power. It has happened several times before and each time everything turned out quite all right. If scientific induction is any guide to the future, we shan't have long to wait.

GROUCHO: Induction, schminduction, not to mention horsefeathers.

RUSSELL: You have a good point there, Mr. Marx. As David Hume showed two hundred years ago, the only warrant for the use of the inductive principle of inference is the inductive principle itself, a clearly circular affair and not really very reassuring.

GROUCHO: Circular affairs are never reassuring. Did I ever tell you about my brother, sister-in-law, and George Fenniman?

RUSSELL: I don't believe you have, though I suspect you may not be referring to the same sort of circle.

GROUCHO: You're right, Lordie. I was talking more about a triangle, and not a cute triangle either. An obtuse, obscene one.

RUSSELL: Well, Mr. Marx, I know something about the latter as well. There was, you may recall, a considerable brouhaha made about my appointment to a chair at the City College of New York around 1940. They objected to my views on sex and free love.

GROUCHO: And for that they wanted to give you the chair?

RUSSELL: The authorities, bowing to intense pressure, withdrew their offer and I did not join the faculty.

GROUCHO: Well, don't worry about it. I certainly wouldn't

want to join any organization that would be willing to have me as a member.

RUSSELL: That's a paradox.

GROUCHO: Yeah, Goldberg and Rubin, a pair o' docs up in the Bronx.

RUSSELL: I meant my sets paradox.

GROUCHO: Oh, your sex pair o' docs. Masters and Johnson, no doubt. It's odd a great philosopher like you having problems like that.

RUSSELL: I was alluding to the set M of all sets that do not contain themselves as members. If M is a member of itself, it shouldn't be. If M isn't a member of itself, it should be.

GROUCHO: Things are hard all over. Enough of this sleazy talk though. (*Stops and listens.*) Hey, they're tapping a message on the girders. Some sort of a code, Bertie.

RUSSELL: (*Giggles*) Perhaps, Mr. Marx, we should term the girder code a Godel code in honor of the eminent Austrian logician Kurt Godel.

GROUCHO: Whatever. Be the first contestant to guess the secret code and win $100.

RUSSELL: I shall try to translate it. (*He listens intently to the tapping.*) It says "This message is . . . This message is . . ."

GROUCHO: Hurry and unlox the Godels, Bertie boy, and st-st-stop with the st-st-stuttering. The whole elevator shaft is beginning to shake. Get me out of this ridiculous column.

RUSSELL: The tapping is causing the girders to resonate. "This message is . . .

A LOUD EXPLOSION.
THE ELEVATOR OSCILLATES SPASMODICALLY
UP AND DOWN.

RUSSELL: ". . . is false. This message is false." The statement
as well as this elevator is ungrounded. If the message is
true, then by what it says it must be false. On the other
hand if it's false, then what it says must be true. I'm
afraid that the message has violated the logic barrier.

GROUCHO: Don't be afraid of that. I've been doing it all my
life. It makes for some ups and downs and vice versa, but
as my brother Harpo never tired of not saying: Why a
duck?

chapter two

LOGIC

There are no more basic principles of logic than the law of noncontradiction and the law of the excluded middle, and hence there is no better place to start the study of logic than with them. The law of noncontradiction states: "It is not the case that A and not A"—or, as Aristotle phrases it: "The same attribute cannot at the same time belong and not belong to the same subject and in the same respect." The law of the excluded middle states: "Either A or not A"—or, to state a specific instance of the law: "Either Wittgenstein was a redhead or he was not." (Symbolically, using $\sim$ for "not," $\wedge$ for "and," $\vee$ for "or," and parentheses ( ) to indicate that the statement within them is to be taken as a whole, the law of the excluded middle can be expressed as $A \vee \sim A$ and the law of noncontradiction as $\sim (A \wedge \sim A)$.)

Even such basic principles as these can cause problems, however, if they're used uncritically. Consider, for example, the law of the excluded middle. Below are three uses of the principle: one, unexceptionable and vaguely humorous; the second, puzzling and distinctly misleading; and the third, straightforward except to a minority.

The first is a story due to Leo Rosten (1968) that tells of a famous rabbi-logician who was so wise that he could analyze any situation no matter how complex. His students wondered, though, if his reasoning power could withstand a bout of drinking—so these respectful yet curious students induced him during a feast to drink enough wine to make him quite tipsy. When he fell asleep, they carried him to the cemetery and laid him on the ground behind a tombstone. They then hid themselves and awaited his analysis of the situation.

When he awoke they were most impressed by his Talmudic use of the law of the excluded middle: "Either I'm alive or I'm not. If I'm living, then what am I doing here? And if I'm dead, then why do I want to go to the bathroom?"

The second instance deals with future events. If it's true now that I shall do a certain thing next Tuesday, let's say fall off a horse, then no matter how I resist doing so, no matter what precautions I take, when Tuesday comes, I shall fall off a horse. On the other hand, if it's false now that next Tuesday I shall fall off a horse, then no matter what efforts I make to do so, no matter how recklessly I ride, I shall not fall off a horse that day. Yet that the prediction is either true or false is a necessary truth—the law of the excluded middle. It seems to follow that what shall happen next Tuesday is already fixed, that in fact not just one event next Tuesday but the entire future is somehow decided, logically preordained.

The problem with the above is not the law of the excluded middle, but the meaning, or rather meaninglessness, of statements of the form "It is true now that some specified event will happen."

Interestingly, a small minority of mathematicians deny that the law of the excluded middle is a law of logic. They object to statements such as "Either there is a string of eight consecutive 5s somewhere in the decimal expansion of π or there is not." Since there is lacking both a constructive proof of the existence of this string of 5s *and* a constructive proof of its nonexistence, the intuitionists and constructivists do not count the above third and most common use of the law of the excluded middle as being valid. For them, truth is a matter of constructive provability.

For quite different reasons, some quantum physicists also reject the applicability of the law of the excluded middle in some contexts. In fact ever since J. Lukasiewicz, an influential Polish logician of the 1920s, initiated the formal study of three-valued logics—true, false, undetermined (indeterminate, intermediate)—it has been an object of persistent, though limited, interest. In more benighted times, classical logicians who accepted the law of the excluded middle sometimes mocked those who did not with the following: "Did you hear the one about the Polish logician? He thought there were three truth values."

The moral of these stories is simply that even such a basic principle of logic can be misapplied, can be controversial. Logic is the most important theoretical tool we possess but, as with all tools, one must know how and when to use it. We want to avoid the sad fate of that proverbial tribe of Indians who, being experts on the theoretical properties of arrows (vectors), simultaneously fired arrows northward and westward whenever they spotted a bear to the northwest, or two arrows northward and one eastward when they spotted a bear north-northeast of them.

⟳

"If I had a horse, I'd horsewhip you!"

—*Groucho Marx*

Conditional statements can be tricky, even straightforward ones like the following.

*: "If George is hungry, then Martha is hungry." It's clear, I hope, that if * is true and George is hungry, then Martha is also hungry. It's equally clear that if George is hungry and Martha is not, then * is false. What if Martha is hungry and it's unknown whether or not George is? In almost all mathematical (and many logical) contexts, the convention is that in this case * is true. What if George is not hungry and it's unknown whether or not Martha is? Again, the convention most useful in mathematics and logic is that * is true.

To summarize: In mathematical, logical, and many everyday contexts any sentence having the form "If P, then Q" or "P implies Q," or, symbolically, "$P \rightarrow Q$," is (1) true whenever Q is true whether P is true or not, (2) true whenever P is false whether Q is true or not, and (3) false only when P is true and Q is false.

⟳

The following two stories are relevant and illustrative:

Bertrand Russell was discussing conditional statements of the above type and maintaining that a false statement implies anything and everything. A skeptical philosopher questioned him, "You mean that if 2 + 2 = 5, then you are the Pope?" Russell answered affirmatively and supplied the following amusing "proof":

"If we're assuming 2 + 2 = 5, then certainly you'll agree that subtracting 2 from each side of the equation gives us 2 = 3. Transposing, we have 3 = 2, and subtracting 1 from each side of the equation gives us 2 = 1. Thus since the Pope and I are two people and 2 = 1, then the Pope and I are one. Hence I'm the Pope."

LOGICIAN: So you see, class, anything follows from a false statement.

STUDENT: I'm afraid I'm lost.

LOGICIAN: It's really quite simple. Are you sure you don't understand?

STUDENT: All I'm sure of is that if I understood that stuff, then I'd be a monkey's uncle.

LOGICIAN: You're right there. (*Laughs*)

STUDENT: Why are you laughing?

LOGICIAN: You wouldn't understand.

STUDENT: Anyway, doc, if you're interested, we're having a party tonight.

LOGICIAN: And if I'm not interested?

STUDENT: What?

LOGICIAN: Thanks anyway, but I'm busy.

Nonmathematical contexts in which the foregoing analysis of if-then statements does not hold are not hard to find. Two statements that are false despite the falsity of their if-clauses are:

(1) If one were to place that nail in that glass of water, it would dissolve.

(2) If Harpo Marx had spoken in any of the Marx brothers' movies, World War II would have been averted.

The truth of such so-called subjunctive and counterfactual conditional statements does not, as in the case of mathematical conditional statements "If P, then Q," depend only on the truth or falsity of P and Q. Rather, it depends on the lawlike relationship that may or may not exist between P and Q.

Still, there are many uses of the mathematical conditional outside of logic and mathematics. If someone says "If it's raining, I'm going to punch you, and if it's not raining, I'm going to punch you," you can be sure that that person is using the mathematical conditional and that he intends to punch you.

Imagine a very rich superscientist who claims to have the power to predict with great accuracy which of two alternatives a person will choose. Imagine further that this scientist, let's call him Dr. Who, sets up a booth at a big World's Fair somewhere in the Midwest to demonstrate his abilities. Dr. Who explains that he tests people by using two boxes: box A is transparent and contains $1,000, whereas box B is opaque and contains either nothing or $1,000,000. Dr. Who tells each person that he or she can choose to take the contents of box B alone, or the contents of both box A and box B. However—and this is important—if, before the person chooses, Dr. Who believes that he or she will take the contents of both boxes, he leaves box B empty. On the other hand, if, before the person chooses, Dr. Who believes that he or she will take only the contents of box B, he places $1,000,000 in box B. Witnesses can verify afterward whether or not the $1,000,000 was placed in box B.

George and Martha are at the World's Fair and see for themselves that when a person chooses to take the contents of both boxes, 95 percent of the time box B is empty and the person gets to keep only the $1,000 in box A. They also note that when a person chooses to take the contents of box B alone, 95 percent of the time it contains $1,000,000, making the person an instant millionaire.

Now it's Martha's turn. Dr. Who examines her carefully, prepares the boxes, places them in front of her, and goes on to George. Impressed by these demonstrations, Martha chooses the contents of box B alone, hoping that Dr. Who accurately assessed her state of mind.

Next it's George's turn. Dr. Who examines him carefully, prepares the boxes, places them in front of him, and goes on to the next person. George reasons that since Dr. Who has already gone, and since the $1,000,000 either has or hasn't been placed in box B already, he may as well choose the contents of both boxes, thereby insuring himself at least $1,000 and possibly $1,001,000.

Finally it's your turn. Dr. Who has already examined you. What choice will you make? (Selling your right to make the choice to someone else for $500,000 is cheating.) The reaction to this paradox, due to the physicist William Newcombe and made well-known by the philosopher Robert Nozick, is intimately connected with one's attitude toward free will, determinism, and money.

Another, better-known story of choice and "higher powers" is due to Blaise Pascal, whose argument for becoming a Christian takes the form of a wager. If, according to Pascal (1966), one elects to believe in Christian doctrine, then if this doctrine is false, one loses nothing—whereas if it's true, one gains everlasting life in heaven. On the other hand, if one elects not to believe in Christian doctrine, then if this doctrine is false, one loses nothing—whereas if it's true, one suffers unrelenting punishment in hell. Only if one already believes in Christian doctrine, as Pascal did, does this argument have any persuasive power. The argument, of course, has nothing to do with Christianity, and could be used by any other religion (or cult) to rationalize other already-existing beliefs.

Everyone would agree that "*A* is false," "*B* is false," "*C* is false," and "One of *A, B,* or *C* is true" are, taken together, an inconsistent set of statements. Yet consider a lottery having 1,000,000 entrants, including you: you don't believe ticket 1 will win, nor ticket 2, nor ticket 3, . . . nor ticket 1,000,000; still, you do believe that *some* ticket will win.

A similar tale can be told about your attitude toward what you read in the newspaper. Unless you have first-hand knowledge of a story, you tend to believe each individual item you read in the paper; still, you also believe that some (many) of the items are false. An understanding of this seeming inconsistency is what prompts celebrities to sue sensationalist tabloids.

Get your tickets for the million-dollar lottery right here. Ten cents apiece, three for a quarter, five for a dollar. Step right up and win a million dollars—a dollar a year for a million years.

Sillygisms

People underestimate the extent to which play enters into any serious intellectual endeavor. Doing something for the what-if fun of it frees one from the shackles of goal-directed plodding and sometimes leads to otherwise unlikely new insights. (And if it doesn't, so what?) This tradition of intellectual play is very old. Even Plato was not above constructing silly "arguments" for his protagonists. A well-known example in the dialogue *Euthydemus* is the exchange between Dionysodorus and Ctessipus:

DIONYSODORUS: You say you have a dog?

CTESSIPUS: Yes, a villain of one.

DIONYSODORUS: And he has puppies.

CTESSIPUS: Yes, and they are very like himself.

DIONYSODORUS: And the dog is the father of them.

CTESSIPUS: Yes, I certainly saw him and the mother of the puppies come together.

DIONYSODORUS: And he is not yours.

CTESSIPUS: To be sure he is.

DIONYSODORUS: Then he is a father, and he is yours; ergo, he is your father, and the puppies are your brothers.

The argument seems silly, yet one with the same grammatical form seems unobjectionable: Fido is a dog. Fido is yours. Therefore Fido is your dog.

An argument in a different vein is due to Raymond

Smullyan, who is some logician:

> Some cars rattle.
> My car really is some car.
> So no wonder my car rattles.

Getting back to dogs, why is one of the following arguments valid and the other not?

> A dog needs water to survive.
> Therefore my dog Linocera needs water to survive.

> A dog is barking in the backyard.
> Therefore my dog Ginger is barking in the backyard.

Perhaps surprisingly, the following two arguments are valid.

> Everybody loves a lover.
> George doesn't love himself.
> Therefore George doesn't love Martha.

> Either everyone is a lover or some people are not lovers.
> If everyone is a lover, Waldo certainly is a lover.
> If everyone isn't, then there is at least one nonlover. Call her Myrtle.

Therefore if Myrtle is a lover, everyone is.

The validity of the first argument depends on two facts: (1) A person is a lover if that person loves anyone, himself included; (2) "If P, then Q" is true exactly when "If not Q, then not P" is. The validity of the second depends on the conditions under which "If P, then Q" is true.

36 inches = 1 yard
So 9 inches = $\frac{1}{4}$ yard
So $\sqrt{9 \text{ inches}}$ = $\sqrt{\frac{1}{4} \text{ yard}}$
Therefore 3 inches = $\frac{1}{2}$ yard

The temperature is 93.
The temperature will rise this afternoon.
Therefore 93 will rise this afternoon.

Most common manipulations of statements involving logical constants such as "all," "some," and "not" are intuitively clear. The negation of "Everyone is bald" (symbolically, $\forall x B(x)$—for all x, x is bald) is not "No one is bald" but rather "Someone is not bald" (symbolically, $\exists x \sim B(x)$—there exists an x such that x is not bald) or, more longwindedly, "It is not the case that everyone is bald."

Sometimes, though, the meanings of these constants are not so clear. "You can fool some of the people all the time" is, at second glance, ambiguous: it has two contrary interpretations. It can mean that there are certain gullible bumpkins, say George and Martha, who are bamboozled every day of the week; or it can mean that on any given day you choose, you'll always be able to fool some people—say, George on Monday, Martha and Waldo on Tuesday, Waldo and Myrtle on Wednesday, and so on. Using $P(x)$ for "x is a person," $T(y)$ for "y is a time," and $F(x,y)$ for "you can fool x at y," the first interpretation can be rendered symbolically as $\exists x \forall y \ (P(x) \land T(y) \rightarrow F(x,y))$, while the second is $\exists x \forall y \ (P(x) \land T(y) \rightarrow F(x,y))$.

An example of a less common logical word subject to similar ambiguities is the word "most":

*: "Most of the people there had read most of the books discussed."

Does * mean that more than half of the people there had read more than half of the books discussed, or does it mean that more than half of the books discussed had been read by more than half of the people there? That these interpretations differ can be seen if, for example, we assume that there were only three people there and only three books were discussed. The first interpretation corresponds to figure 1, and the second to figure 2.

Figure 1	Figure 2

George — War and Peace   George — War and Peace
Martha — Tropic of Cancer   Martha — Tropic of Cancer
Waldo — Principia   Waldo — Principia
         Mathematica   Mathematica

Even the simple logical word "is," maybe even *especially* the word "is," can be interpreted in everyday contexts in quite different ways. Consider the four statements below:

(1) George is Mr. Kyriacopoulos.

(2) George is anxious.

(3) Man is anxious.

(4) There is an anxious man here.

The first "is" is the "is" of identity: $g = k$. The second "is" is the "is" of predication: $A(g)$—g has the property A. The third "is" is the "is" of membership: $\forall x\, [M(x) \rightarrow A(x)]$. The fourth "is" is existential: $\exists x\, M(x)$.

It's odd that logical acuity, rather than helping one to clarify statements, often reveals hidden ambiguities within them. Instead of leading one to form more conclusions, it makes clear that fewer conclusions are justified. Bertrand Russell once observed (1956) that the keener one's sense of logical deduction, the less often one makes hard and fast inferences.

However, only for deductive inferences is this the case. Most "common sense" inferences do not depend on explicitly formulated knowledge or on any precise rules, but rather on a kind of tacit and natural process that is hard to describe. Thus mathematicians, accustomed as they are to clearly formulated principles and rigidly interpreted rules, sometimes have difficulty with "common sense" and often have a sense of humor characterized by the overly literal interpretation of terms and phrases.

⊚

MINISTER OF WAR: That's the last straw! I resign. I wash my hands of the whole business.

FIREFLY (GROUCHO): A good idea. You can wash your neck too.

⊚

GEORGE: In what two sports do face-offs occur?

WALDO: Ice hockey and . . . I give up. What's the other one?

GEORGE: Leper boxing.

⊚

No Parking signs often indicate "Violators will be towed," yet I've never yet seen a tow truck dragging anyone down the street. Trash cans that warn "Keep litter in its place" are also amusing if taken literally (litterly?): if something is litter, its place by definition, it would seem, is the ground.

⊚

Finally, consider this last sillygism, which leads into the topic of the next section:

This argument is helopita.
Nothing that's helopita is steedibeep.
Everything that's valid is steedibeep.
Therefore this argument is not valid.

Yet it is.

The Titl of This Section
Contains Three Erors

The study of self-referential statements dates back to Stoic logicians of the fourth and fifth century B.C. The oldest, best-known such paradox concerned Epimenides the Cretan, who stated: "All Cretans are liars." The crux of this so-called liar paradox is clearer if we simplify his statement to "I am lying" or, better yet, "This sentence is false." (Douglas Hofstadter writes of a more recent version, that of Nixonides the Cretin, who in 1974 stated: "This sentence is inoperative" [Hofstadter].)

Let us give the label Q to "This sentence is false." Now we notice that, if Q is true, then by what it says it must be false. On the other hand, if Q is false, then what it says is true, and Q must then be true. Hence, Q is true if and only if it's false.

The statement Q and variants of it are intimately connected with some of the deepest and most important ideas in logic and philosophy, and possibly even with consciousness itself. Despite this they are dismissed by most people as silly diversions, suitable only for amusing logicians and other useless people. I must admit that on Tuesdays I feel the same way, and since this is Tuesday, I will demonstrate how a relative of Q can be used to prove the existence of God. Consider the box below, which contains two sentences:

> 1. God exists.
>
> 2. Both of these sentences are false.

The second sentence is either true or false. If it's true, then both sentences are false. In particular, the second sentence is false; but the only way the second sentence can be false is for the first sentence to be true. Thus in this case God exists. On the other hand, if we assume directly that the second sentence is false, then again we must note that the only way for this to be is for the first sentence to be true. Thus in this case also God exists. Hence God exists.

Of course, in a similar way we can demonstrate that God had a hangnail, or that he doesn't exist, or that Ludwig Wittgenstein was in love with Mae West.

A related trick can be pulled with the following statement:

*: "If this statement is true, then God exists." If * is true, then it's true. On the other hand, if * is false, then the antecedent (the if clause) of * is false, which ensures that * itself is true. (Remember that a conditional if-then statement is true if the antecedent is false.) Thus in both cases * is true. Thus * is true, and hence the antecedent of * is true. Together these two facts ensure that the consequent (the then clause of *) is true. Hence God exists.

Even nonparadoxical statements can, if the situation is appropriate, be combined to yield a paradox. If Socrates were to say "What Plato just said is false," there would usually be nothing strange about the utterance. But if Plato were previously to have said "What Socrates says next is true," we would have a paradox.

There is an ancient story about the Sophist philosopher Protagoras, who agreed to instruct Euathlus in rhetoric so the latter could practice law. Euathlus in turn agreed to pay Protagoras his fee only after winning his first case. However, Euathlus chose not to practice law upon completing his training, and so Protagoras sued him for his fee. Protagoras maintained that he should be paid no matter what: he argued that if he won the case, he should be paid by order of the court; while if he lost, he should be paid by the terms of his agreement with Euathlus. Euathlus, who had learned something from his study with Protagoras, maintained that he should *not* pay no matter what: he argued that if he won the case, he should not pay by order of the court; while if he lost, he should not pay by the terms of his agreement with Protagoras.

Breakfast Scene

Big, brawny woman, hair in pincurls, wearing a torn bathrobe, to her scrawny, bald husband sitting in his underwear: "I want you to dominate me, to make me feel like a real woman."

As the last story might suggest, there is a close connection between these paradoxes and certain "double-bind" situations. The command "Be spontaneous" is another simple example, as is the injunction "Whatever you do, don't let the image of a sweet red watermelon covered with thick yellow mustard enter your mind." Unfortunately (or fortunately), most such situations that require contradictory behavior are more complex and somewhat more disguised—so disguised, in fact, that most people, especially those who consider self-referential paradoxes frivolous, do not realize that these situations pervade their lives. Situations, incidents, conversations are often so complexly textured that it's not uncommon for some aspect of these communications to "say" of itself that it's not true.

Consider the act of telling a joke. W. F. Fry and Gregory Bateson have shown that when someone tells a joke, there is usually some kind of behavioral cue—a different voice inflection, an arched eyebrow or wink, the use of a dialect, a mock serious tone, even the explicit clause, "Have you heard the one about. . . ." These cues, or metacues, qualify what is being said and function as a kind of nonverbal liar paradox. They say, in effect, "This whole business is false, unreal, not to be taken seriously; it's only joking." In fact, all dramatic performance—all art, even—has these two aspects: the content, and the frame or setting, which sets it apart from non-art and which says of itself, "This is not an everyday sort of communication. This is unreal."

A man, smiling and holding a small tree branch above his head, informs a bank teller, "This is a stick-up."

Consider a country whose laws allow nonresident mayors—that is, some mayors live in the cities they govern, while others do not. A reform-minded dictator comes to power and orders that all the nonresident mayors, and only the nonresident mayors, live in one place—call it city C. City C now requires a mayor. Where shall the mayor of city C reside?

This story is a popularization of Russell's paradox, whose derivation is similar. We first note that some sets contain themselves as members. The set of all things mentioned on this page is mentioned on this page, and thus contains itself. Likewise, the set of all those sets with more than seven members itself contains more than seven members, and is thus a member of itself. Most naturally occurring sets do not contain themselves as members. The set of hairs on my head is not itself a hair on my head, and thus is not a member of itself. Similarly, the set of odd numbers is not itself an odd number, and thus does not contain itself as a member.

Dividing the set of all sets into two non-overlapping sets, let us denote by M the set of all those sets that do contain themselves as members, and by N the set of all those sets that do not contain themselves as members. In other words, for any set x, if x is a member of M, then x is a member of itself; and if x is a member of itself, then x is a member of

M. On the other hand, for any set *x*, if *x* is a member of *N*, then *x* is not a member of itself; and if *x* is not a member of itself, then *x* is a member of *N*.

Now we may ask whether *N* is a member of itself or not. (Compare this question with "Where does the mayor of city *C* live?") If *N* is a member of itself, then by definition of *N*, *N* is not a member of itself. But if *N* is not a member of itself, then by definition of *N*, *N* is a member of itself. Thus *N* is a member of itself if and only if it is not a member of itself. This contradiction constitutes Russell's paradox.

Robert Benchley once remarked: "There may be said to be two classes of people in the world: those who constantly divide the people of the world into two classes and those who do not." He should have added paradoxically that he belongs to the latter class.

Recall Groucho Marx's quip that he would never join any club that would be willing to have him as a member.

The following is a true story: A well-known philosopher was delivering a talk on linguistics and had just stated that the double negative construction has a positive meaning in some natural languages and a (very) negative meaning in others. He went on to observe, however, that in no language was it the case that a double positive construction has a negative meaning. To this, Sidney Morgenbesser, another well-known philosopher, who was sitting in the rear of the lecture room, responded with a jeering "Yeah, yeah."

V: "This sentence is true." *V* is a little odd, but not paradoxical. A variant of it, due to David Moser, reads: "This sentence is a !!! premature punctuator." "This sentence is graduellement changeant en français" is another.

Russell's resolution of the paradox is to restrict the notion of set to a well-defined collection of already-existing sets. In his famous theory of types, he classifed sets according to their type or level, thereby creating a set-theoretic hierarchy. On the lowest level, type 1, are individual objects. On the next level, type 2, are sets of type 1 objects. On the next level, type 3, are sets of type 1 or type 2 objects or sets; and so on. The elements of type n sets are sets of type $(n - 1)$ or lower. In this way Russell's paradox is avoided, since a set can be a member only of a set of a higher type and not of itself. Informal uses of this idea are commonplace: characters in cartoons, television, and movies are forever saying things like they're anxious about their anxieties, bored with their boredom, or tired of being tired.

The liar paradox can be avoided in a similar way. Applying a hierarchical solution to Epimenides' utterance requires that "All Cretans are liars" be assigned a higher type than other statements made by Cretans. We make a distinction between type I statements (usually called first-order statements), which do not refer to other statements; second-order statements, which refer to first-order statements; third-order statements, which refer to first- and second-order statements; and so on. Thus if Epimenides the Cretan states that all statements made by him are false, he is to be understood as making a statement that does not apply to itself; he is making a statement of an order higher than that of his other statements. In this way the self-negation of the liar paradox is prevented.

The following three statements are first-, second-, and third-order statements, respectively:

(1) Wittgenstein was bald.

(2) Statement (1) is false.

(3) Statement (2) is true.

This hierarchical notion of truth for statements has been extensively developed by the logician Alfred Tarski and others. It is not, however, the only way to handle these questions. An alternative approach, due to the philosopher Saul Kripke, takes statements like Q—"This sentence is false"—to be "ungrounded," there being no ground, no first-order statement, from which to build up to the truth or falsity of Q. Statements in the Kripke formulation are not (as in the Tarski account) assigned fixed levels or orders, but attain their level naturally depending on what other statements have been made and the facts of the situation. (Cf. the Plato-Socrates story above.) The truth or falsity of these statements is determined in a gradual, inductive manner, with not every sentence receiving a truth value (e.g., Q). Self-referential statements are allowed, though, and can, under appropriate circumstances, receive a truth value.

A literal interpretation of "self-addressed stamped envelope" is a stamped envelope addressed to itself. Ideal for instant delivery.

"You always overreact. You never ever respond moderately. If you shrug one more time, I'm going to scream."

AIXƎ⅃SYD

Language and Metalanguage: Do You Get It?

Implicit in the discussion of language levels in Russell's theory of types is a very general and extremely important distinction in logic and philosophy, that between object language and metalanguage. Object-level statements are statements within a (usually) formal system that is the object of study. Examples are:

(1) $A \wedge B \rightarrow {\sim} C \vee B$

    (If A and B, then either not C or B)

(2) $\forall x \exists y P(x,y)$

    (For all x there is a y such that x bears relation P to y)

(3) $p \mid (x^2 - 1) \rightarrow p \mid x$

    (If p divides $x^2 - 1$, then p divides x)

Metalevel statements are statements about the formal system, or about the object-level statements within it. Examples are:

(1) Statement Q has two different interpretations.

(2) Statement S is true but not provable.

(3) P, S, and Q are inconsistent.

If one is studying Japanese grammar, Japanese is the object language and English is the metalanguage.

⟨◎⟩|

Groucho says to an acquaintance, "Did you hear the one about the organization for people whose IQs are in the bottom 2 percent? I was just looking at its newsletter called DENSA." Groucho stops, bends closer, and says casually, "Do you get it? Do you get it?" The acquaintance, thinking he's being asked if he understands the joke, answers "Yes," to which Groucho rejoins, "I'm surprised. I thought you were a little brighter than that."

Risking a charge of pedantry, I'll note that Groucho's acquaintance misinterprets "Do you get it?" as a metalevel question about the joke, and not an object-level question that is part of the joke.

The tortoise in Lewis Carroll's "What the Tortoise Said to Achilles" (excerpted below) makes the opposite mistake. He confuses the metalevel rule C with the object-level statements A, B, and Z. C needs no further rule to explain when and how it applies, yet the tortoise insists on a meta-metalevel rule D to do just that, and then a meta-metametalevel rule E to explain when and how D applies, and so on. The tortoise just isn't playing the logic game.

"That beautiful First Proposition by Euclid!" the Tortoise murmured dreamily. "You admire Euclid?"

"Passionately! So far, at least, as one can admire a treatise that won't be published for some centuries to come!"

Well, now, let's take a little bit of the argument in that First Proposition—just two steps, and the conclusion drawn from them. Kindly enter them in your note-book. And in order to refer to them conveniently, let's call them A, B, and Z:

A) Things that are equal to the same are equal to each other.

B) The two sides of this Triangle are things that are equal to the same.

Z) The two sides of this Triangle are equal to each other. . . .

"I'm to force you to accept Z, am I?" Achilles said musingly. "And your present position is that you accept A and B, but you *don't* accept the Hypothetical—"

"Let's call it C," said the Tortoise.

"—but you *don't* accept C) If A and B are true, Z must be true."

"That is my present position," said the Tortoise.

"Then I must ask you to accept C."

"I'll do so," said the Tortoise, "as soon as you've entered it in that note-book of yours. What else have you got in it?"

"Only a few memoranda," said Achilles, nervously fluttering the leaves: "A few memoranda of—of the battles in which I have distinguished myself!"

"Plenty of blank leaves, I see!" the Tortoise cheerily remarked. "We shall need them *all*!" (Achilles shuddered.) "Now write as I dictate:—

A) Things that are equal to the same are equal to
   each other.

B) The two sides of this Triangle are things that
   are equal to the same.

C) If A and B are true, ζ must be true.

Z) The two sides of this Triangle are equal to each
   other."

"You should call it D, not ζ," said Achilles. "It
comes *next* to the other three. If you accept A and B
and C, you *must* accept ζ."

"And why *must* I?"

"Because it follows *logically* from them. If A and B
and C are true, ζ *must* be true. You don't dispute *that*,
I imagine?"

"If A and B and C are true, ζ *must* be true," the
Tortoise thoughtfully repeated. "That's *another*
Hypothetical, isn't it? And, if I failed to see its truth,
I might accept A and B and C, and *still* not accept ζ,
mightn't I?"

"You might," the candid hero admitted; "though
such obtuseness would certainly be phenomenal.
Still, the event is *possible*. So I must ask you to grant
one more Hypothetical."

"Very good. I'm quite willing to grant it, as soon
as you've written it down. We will call it D) If A and B
and C are true, ζ *must* be true. Have you entered that
in your note-book?"

"I *have*!" Achilles joyfully exclaimed, as he ran the
pencil into its sheath. "And at last we've got to the

end of this ideal racecourse! Now that you accept *A*
and *B* and *C* and *D*, *of course* you accept *Z*."

"Do I?" said the Tortoise innocently. "Let's make
that quite clear. I accept *A* and *B* and *C* and *D*.
Suppose I *still* refused to accept *Z*?"

"Then Logic would take you by the throat, and
force you to do it!" Achilles triumphantly replied.
"Logic would tell you 'You can't help yourself. Now
that you've accepted *A* and *B* and *C* and *D*, you must
accept *Z*!' So you've no choice, you see."

"Whatever *Logic* is good enough to tell me is worth
writing down," said the Tortoise. "So enter it in your
book, please. We will call it E) If *A* and *B* and *C* and *D*
are true, *Z* must be true. Until I've granted *that*, of
course I needn't grant *Z*. So it's quite a *necessary* step,
you see?"

"I see," said Achilles; and there was a touch of
sadness in his tone.   (*Carroll*)

Godel's famous incompleteness (meta)theorem also
depends crucially on the object-metalevel distinction.
Godel considered a simple formal system containing the
basic axioms of the arithmetic of whole numbers. He
methodically assigned each object-level statement a unique
code number; he also assigned a code number to each proof
of an object-level statement. By means of this coding,
object-level statements about numbers can also be under-

stood as expressing metalevel statements about the system, or about individual object-level statements. If one is careful and clever, one can find a statement G that is true if and only if it is unprovable.

Loosely speaking, we note that this object-level statement about whole numbers says of itself, via the numerical coding, that it is not provable. If the axioms are all true and the system is consistent, it is possible to conclude that such a statement G (about whole numbers, remember) is neither provable nor disprovable from the axioms—that it is independent of them. The same idea can be extended to show that any formal system is incapable of proving some truths and thus that no formal system can prove all truths.

The following old joke has something of the same flavor: A new prisoner was puzzled because his fellow inmates laughed whenever one of them called out a number. He was told that the numbers were a code for certain jokes, which thus did not need to be repeated verbatim. Intrigued, the new prisoner called out "63" and was greeted by total silence. Later his cellmate explained that everything depends on how the joke is told.

This metajoke can, I suppose, itself be assigned a code number and. . . .

Comedians, after a joke has failed, often follow it with a self-deprecating comment on the joke, thereby salvaging at least a metajoke. So do authors.

This sentence has three erors. As in the titl of a previous section, one of these erors is of a quite different type (level) than the other two.

Related to the language-metalanguage distinction—in fact, a special case of it—is the use-mention distinction. To illustrate, the first pair of sentences below uses the words "laughing" and "Daniel," while the second pair merely mentions them:

(1) Leah was laughing at the Smurf cartoons.

(2) Daniel loves lawnmowers.

(1') "Laughing" contains eight letters.

(2') "Daniel" is a boy's name.

That the distinction can be grasped by a five-year-old is demonstrated by the following true story: A colleague of mine, working on a paper at home in his study, was disturbed repeatedly by his five-year-old son's use of the expletive "shit" in the next room whenever his blocks collapsed. My colleague firmly warned him not to use the word anymore. Returning to his study he heard his son say it again, whereupon he whirled around and barged into his son's room only to hear him continue defensively ". . . is a bad word. 'Shit' is a bad word. Right, Daddy?"

Failure to distinguish between using and mentioning can also lead to arguments like the following concerning former-President and Mrs. Ford:

Betty loves Ford.
Ford is a four-letter word.
Therefore Betty loves a four-letter word.

i should begin with a capital letter.

Meaning, Reference, and Dora Black's First Husband

Meaning, reference, names, and descriptions: these notions are at the heart of many disputes in philosophical logic. Whatever the resolution of these, some well-known puzzles will have to be solved (or dissolved).

The meaning of "meaning" is difficult if not impossible to formulate in full generality. One clear though wrongheadedly narrow account was provided by the early logical positivists, who identified the meaning of a proposition with its method of verification—that is, with whatever observations point to its truth. Using this so-called verifiability principle, they thought they had managed to do away with metaphysics, theology, and ethics, whose propositions are seemingly unverifiable. Hence we have the boring sterility of the logical positivists' program. One major problem with this theory of meaning, of course, is that the verifiability principle itself is embarrassingly not verifiable. "Reference" too is a difficult notion to pin down. Still, whatever these terms mean we can say something about the connection between them.

Two terms or expressions can refer to the same entity (or set of entities) though they differ in meaning. To use a classical example, due to the logician Gottlob Frege, we note that "the morning star" and "the evening star" certainly differ in meaning, or, as is sometimes said, in intension (with an *s*). It required an empirical discovery to realize that both those expressions have the same referent (i.e., both refer to the same entity)—the planet Venus. Similarly, "the younger coauthor of *Principia Mathematica*" and the "first husband of Dora Black," though they do not mean the same thing, both refer to Bertrand Russell. The term "renates" means "animals with kidneys," while "cordates" means "animals with hearts"; yet they both refer to the same collection of animals, since it so happens that all animals with hearts have kidneys and vice versa.

A term or expression is said to occur *extensionally* in a sentence if replacing it with any other term or expression with the same referent does not change the truth or falsity of the sentence. In many contexts, certainly in all purely mathematical ones, this substituting of equals for equals is extensional and causes no problem. Its use in fact was so obvious to Euclid that he included it in his development of plane geometry as "a common notion" not in need of any further elucidation or proof. If a is greater than x^2 <ms> I, and $a = b$, then b is greater than x^2 <ms> I. Still, outside of mathematics this substitution principle can fail—as the following arguments show:

The president thought that the city of Copenhagen was in Norway.

The city of Copenhagen is the capital of Denmark.

Therefore the president thought the capital of Denmark was in Norway.

It is a mathematical theorem that the number six is greater than the number three.

The number six is the number of men who were husbands of Elizabeth Taylor.

Therefore it is a mathematical theorem that the number of men who were husbands of Elizabeth Taylor is greater than the number three.

Statements or expressions can have a meaning yet lack a referent (at least according to many philosophers' accounts). Russell's celebrated "The present King of France is bald" is an example. Russell takes it to mean "There is a single person who is King of France, and that person is bald"; on this analysis it is meaningful but false.

GEORGE: Peter Pan doesn't exist.

MARTHA: You mean the boy who flies through the air, who does battle with Captain Hook, and whom little children love.

GEORGE: Yes, he doesn't exist.

MARTHA: Who doesn't exist?

GEORGE: Peter Pan.

WALDO: When are you going to pay the balance of this bill, George?

GEORGE: Don't worry. I'll get it to you by the second Tuesday of next week.

Sometimes even more serious difficulties can arise, as when George announces, "My brother is an only child."

Just as two terms can differ in meaning yet have the same referent, a term can have many referents but only one meaning. "My father," for example, when uttered by me refers to my father, whereas when uttered by Waldo, it refers to Waldo's father (surprise). Similarly, words like "you" or "I," "yesterday" or "tomorrow," "here" or "there" refer to different entities depending on when, where, and by whom they're uttered.

A man and an acquaintance of his are walking down a street one afternoon. The man spots his wife and his mistress talking in a café and amusedly remarks, "Imagine a mistress spending the morning with her lover and then having a friendly chat with his wife that afternoon." The acquaintance, a pale, shocked look on his face, responds "How did you find out?"

Two clergymen were discussing the present sad state of sexual morality. "I didn't sleep with my wife before we were married," one clergyman stated self-righteously. "Did you?"

"I'm not sure," said the other. "What was her maiden name?"

Analytic vs. Synthetic, Boole vs. Boyle, and Mathematics vs. Cookery

An analytic truth is one that is true in virtue of the meanings of the words it contains, and a synthetic truth is one that is true in virtue of the way the world is. ("If George is smelly and bald, then he's bald," vs. "If George is smelly, then he's bald." "Bachelors are unmarried men," vs. "Bachelors are lascivious men." "UFOs are flying objects that have not been identified," vs. "UFOs contain little green men.") This distinction is a sprucing up of Immanuel Kant's original one, which in turn derives from similar distinctions due to David Hume and Gottfried Leibniz. Some philosophers, in particular the American W. V. O. Quine, have argued that the distinction is not hard and clear, but rather one of degree or convenience. It is still, even if not absolute and immutable, a useful distinction.

When Molière's pompous doctor announces that the sleeping potion is effective because of its dormitive virtue, he is making an empty, analytic statement, not a factual, synthetic one. The same thing can be said about the White Knight in *Through the Looking Glass* when he describes to Alice the song he wants to sing:

"It's long," said the Knight, "but it's very, *very* beautiful. Everybody that hears me sing it—either it brings tears to their eyes, or else—"

"Or else what?" said Alice, for the Knight had made a sudden pause.

"Or else it doesn't, you know."

Likewise, "A well-told joke will not be funny unless it's well-told," despite the appearance that it's saying something substantial, is simply an analytic truth. It is equivalent to "If a well-told joke is funny, it is well-told."

Conversely, synthetic truths are sometimes mistaken for statements that are analytically true or analytically false. Examples are "The second-place Chicago White Sox took the field wearing white sox," or the perennial "The Holy Roman Empire was not holy, Roman, or imperial." Some of the pronouncements on space and time in Einstein's theory of relativity are also seen now to be synthetic truths and not, as they appeared and still appear to many people, analytic falsehoods.

Even more common are exchanges like the following, in which prejudices are preserved by making them, through redefinition, analytically true:

GEORGE: Scotsmen don't buy jewelry.

MARTHA: But MacGregor just bought fourteen diamond necklaces.

GEORGE: MacGregor is not a true Scotsman, then.

Capitalists (neurotics, Jews, Englishmen, Greeks, blacks, etc.) don't do such and such. But so and so did do such and such. Well then, he's not a real capitalist (neurotic, Jew, Englishman, Greek, black, etc.).

It can be said that the difference between analytic truths and synthetic truths is the difference between an "*o*" and a "*y*," between Boole's laws of logic and Boyle's laws of gases. It is roughly the difference between the formal sciences (mathematics, logic, and linguistics) and the empirical sciences (physics, psychology, and cooking).

Bertrand Russell once wrote:

Pure mathematics consists entirely of such asservations as that, if such and such a proposition is true of *anything*, then such and such another proposition is true of that thing. It is essential not to discuss whether the first proposition is really true, and not to mention what the anything is of which it is supposed to be true. . . . If our hypothesis is about *anything* and not about some one or more particular things, then our deductions constitute mathematics. Thus mathematics may be defined as the subject in which we never know what we are talking about, nor whether what we are saying is true.   *(Russell)*

Though the ubiquity of people who do not know what they're talking about nor whether what they're saying is true may suggest that mathematical talent is widespread, the quote does give a succinct summary of the formal approach to mathematics. Certain axioms, expressed in a formal language, are laid down; precise rules of inference are formulated; and theorems are derived from the axioms by means of the rules of inference. What anything means is (or can be) ignored. In this respect, mathematics can be compared to the game of chess—the axioms to the initial positions, the rules of inference to the rules governing the allowable moves, and theorems to subsequent positions of the pieces. Mathematical truths, in particular Euclidean geometric truths, were thought to be (using Immanuel Kant's terms) synthetic a priori—that is, they were considered to be true because of the way the world is, yet independent of experience. The development of consistent non-Euclidean geometries by Bolyai, Lobachevsky, and Gauss led to the realization, implicit in Russell's quote, that points, lines, and other primitive geometric terms and relations could be taken to be *anything at all* that satisfied the formal axioms containing these terms and relations, and that geometrical theorems were simply any formal statements that followed from the axioms by means of the rules of inference.

Not until (if ever) the terms are given a particular empirical meaning do the notions of truth or falsity become appropriate. As the French mathematician Poincaré once wrote, "What are we to think of the question: Is Euclidean geometry true? It has no meaning. . . . One geometry cannot be more true than another; it can only be more convenient" (Poincaré 1913). This is because mathematics chases not truth (a metalevel notion) but formal consequences (provability, an object-level notion), not "Is this true of the world?" but "Does this follow from that?" Einstein phrased it: "As far as the properties of mathematics refer to reality, they are not certain: and as far as they are certain, they do not refer to reality" (Einstein). Mathematical truths, by and large, are certain because they are analytic; physical truths are not certain because they are synthetic.

George goes to the You-Bet-Your-Life computer dating service to register his requirements (axioms). He wants someone who is white, not very talkative, comfortable in fur, yet disdainful of city life. The computer sends him a polar bear.

Immanuel Kant goes to the Lobachevsky Lumber Company to order a flat board with which to cover his desk. Feeling mischievous, he asks for a surface that satisfies the (first four) axioms of Euclidean plane geometry. The lumber company gives him a saddle-shaped piece of wood.

That *synthetic* scientific laws and facts cannot be determined a priori now seems a commonplace. That this was not always the case is illustrated by the following excerpt from Francesco Sizi wherein he "argues" that, contrary to what his contemporary Galileo had seen through his telescope, Jupiter could have no satellites:

> There are seven windows in the head, two nostrils, two ears, two eyes and a mouth; so in the heavens there are two favorable stars, two unpropitious, two luminaries, and Mercury alone undecided and indifferent. From which and many other similar phenomena of nature such as the seven metals, etc., which it were tedious to enumerate, we gather that

the number of planets is necessarily seven. . . .
Moreover, the satellites are invisible to the naked eye
and therefore can have no influence on the earth
and therefore would be useless and therefore do not
exist.

The German philosopher Hegel, just before the discovery of the asteroid Ceres, also chastised astronomers for
not paying more attention to philosophy—a science that
would, according to Herr Hegel, have shown them that
there could not possibly be more than seven planets.

Miscellany

I'd like to close this chapter with a short collection of logical jokes and stories. The first "joke" consists merely of pairs of phrases, each element of the pair sharing the same grammar yet having a different logic (in a broad, informal sense of "logic"):

"Waldo likes to move his rooks out early," vs. "The MATE-IAC II computer likes to move its rooks out early."

"Going on to infinity," vs. "going on to Paris."

"Honesty compels me," vs. "The IRS compels me."

"The present Czar of Russia is obese," vs. "The present president of the United States is obese."

"Being a baseball player," vs. "being a baseball."

"An alleged murderer," vs. "a vicious murderer."

"Have you stopped beating your husband?" vs. "Have you voted for Megalomeeti yet?"

"If only Pat had a different job," vs. "If only Pat had a different sex."

"Before the world began," vs. "before the Phillies game began."

"Studying for a physics test," vs. "studying for a urine test."

Modern analytic philosophy has sometimes been called linguistic therapy, and philosophers like Wittgenstein, Ryle, and Austin have devoted much effort and analysis to curing some of the "linguistic diseases" lurking all over, in particular in phrases such as the above.

⊚

T.V. SPORTSCASTER: Folks, we're running out of time so I'll have to hurry with the baseball scores. 4 to 2, 6 to 3, 8 to 5, and in a real cliff-hanger, 2 to 1.

⊚

T.V. FOOTBALL COMMENTATOR: These teams really came to play today.

⊚

PEASANT: Is kebab with an "a" or an "o"?
SUFI MASTER: With meat.

⊚

10-YEAR-OLD: Pete and Repeat were walking down the street. Pete fell down. Who was left?
7-YEAR-OLD: Repeat.
10-YEAR-OLD: Pete and Repeat were walking down the street. Pete fell down. Who was left?

⊚

MARTHA: George, in this game cheating is allowed.

A wife laughs at her distraught husband who has a loaded revolver at his temple. "Don't laugh," he tells her. "You're next."

Title of Book: *20 Ways to Regain Your Virginity*

There was once a very brilliant horse who mastered arithmetic, algebra, plane geometry, and trigonometry. When presented with problems in analytic geometry, however, the horse would kick, neigh, and struggle desperately. One just couldn't put Descartes before the horse.

What's a question that contains the word "cantaloupe" for no apparent reason?

COSTELLO: Who's on first?
ABBOT: Yeah. Who is on first, Johnson's on second.
COSTELLO: Who's on first?

ABBOT: Right, Who is. Johnson's on second and Walters is up to bat.

COSTELLO: But who is on first?

WAITER: Would you like white wine or red wine with your dinner?

GEORGE: It doesn't matter. I'm color blind.

Speaking of liquids, the following is an old conundrum: A tablespoon of water is removed from an 8-oz. glass of water, placed in an 8-oz. glass of wine, and the resulting mixture stirred. A tablespoon of the mixture is then removed, placed in the water glass, and stirred. Is there more wine in the water or more water in the wine?

If you're sick of such logical niceties, the last story may please you. Bertrand Russell, in satirizing the empty precision of some philosophers, told of a man who comes to a fork in the road. He asks a philosopher who happens to be loitering there, "Which way to the town of Dresher?" The philosopher responds, "Which one of these two roads here?"

"Yes, yes."

"It's the town of Dresher you're looking for?"

"Yes, yes."

"You want to follow one of these two roads here to the town of Dresher?"

Growing impatient, the man repeats, "Yes, which road leads to Dresher?"

"I don't know."

The philosopher was not a very good one. He should have asked the man if he was sure that one and only one of the two roads really did lead to Dresher, before telling him that he didn't know which. A more significant and disturbing revelation of ignorance begins the next chapter.

chapter three

SCIENCE

Induction, Causality, and Hume's Eggs

WOMAN: Doctor, doctor. You must help me. My husband thinks he's a chicken.

DOCTOR: That's terrible. How long has he thought this way?

WOMAN: As long as I can remember.

DOCTOR: Then why didn't you see me sooner?

WOMAN: I would have, but we needed the eggs.

If the doctor were to answer that he too needed the eggs, we would have something analogous to the problem of induction:

WOMAN: Professor, professor. You must help me. My husband uses an inductive argument to justify the use of inductive arguments.

PROFESSOR HUME: That's terrible. How long has he acted this way?

WOMAN: As long as I can remember.

HUME: Then why didn't you see me sooner?

WOMAN: I would have, but we needed (the conclusions of) the inductive arguments.

HUME: I'm afraid I need them too.

Let me summarize what is usually referred to as Hume's traditional problem of induction or, as Russell once called it, "the scandal of philosophy." We, every day of our lives, confidently use inductive arguments (arguments whose conclusions go beyond, contain more information than, the premises). Why are we so confident that these arguments usually yield true conclusions from true premises? There certainly is no deductive argument that since the sun has risen regularly in the past, it will probably rise tomorrow, or that since stones that have been dropped have always fallen in the past, they will probably fall when dropped in the future. It seems that the only argument for the continuation of these regularities is an inductive one: since these regularities have obtained in the past, they will probably continue into the future. But to try to justify the use of inductive arguments by an inductive argument is clearly circular and begs the question. To put the matter a little crudely, the answer to the question "Why will the future be like the past in certain relevant respects?" is nothing more satisfying than "It will be because past futures have been like past pasts in certain relevant respects." This is helpful only if the future will be like the past—which is the point at issue.

There have been many attempts to clean up "the scandal of philosophy." One way out is just to accept a nonempirical principle of the uniformity (over time) of nature. The problem with this "solution" is that it also begs the question; it is equivalent to what is to be established. It has the advantage, as Russell said in a different context, of "theft over honest toil." Another attempted way out is to note that some inductive arguments are of higher order than others, and to try to make use of this hierarchy (of inductive arguments, meta-inductive arguments, meta-metainductive arguments, and so on) to somehow justify induction. This does not quite work—or rather it works too well, and "justifies" a lot of weird practices.

Charles Saunders Peirce and Hans Reichenbach have advanced a different pragmatic justification of induction, which amounts roughly to this: "Maybe induction does not work, but if anything does, induction will. Maybe there is no order in the universe, but if there is any (on any level), induction will eventually find it (on the next-highest level)." There is some merit to this approach, but there is also a problem with the word "eventually." Finally, there has been an attempt to dissolve the problem by showing that following our commonsense inductive rules is what is meant by rationality, and therefore no further justification is called for.

I wrote in the introduction that philosophy is not a guide to life, a branch of theology or mathematics, or merely a matter of being stoical in the face of adversity. Whatever the resolution of Hume's traditional problem of induction, it beautifully exemplifies the nature of philosophical inquiry. Once Hume enunciated his insight into induction (and causality, which we will treat later), it became impossible for anyone to think about induction in the same way again. No new facts or theorems or prescriptions were offered, just a sometimes scary realization that induction is not what it appears to be in our uncritical daily life, where our focus is usually on the eggs we need.

A Humerous induction:

MARTHA: On all my birthdays up to now I've been less than twenty-five years old. So by induction, on all my birthdays I'll be less than twenty-five years old.

Hume's analysis of the notion of causality is similarly unsettling. According to Hume, when we say "A causes B" we mean nothing more than that A and B are constantly conjoined, that in every instance we've examined, the event A has been followed by the event B. Since it's quite easy to imagine A not being followed by B, the connection between A and B cannot be a necessary or logical one. Causes and effects are discoverable by experience and not by a priori reasoning. Many people—including, of course, Kant—believe this.

Still, there are problems with this view of cause and effect. Consider, for example, scientific laws. Are they merely concise summaries of past "constant conjunctions" of A's and B's, descriptive restatements of A's being followed by B's? What makes it hard to maintain this view is, as Nelson Goodman has noted, that scientific laws, unlike accidental generalizations, seem to support counterfactual conditionals. (A counterfactual conditional is a statement of the form "If A were the case, then B would be the case," when in fact A is *not* the case.)

Thus the scientific law "All objects with a specific density greater than that of water sink when placed in water" supports the counterfactual conditional "If this nail had been placed in the water, it would have sunk." The accidental generalization "All the students in George's Math 5 class are functionally illiterate" does not, however, support the counterfactual conditional "If Martha were placed in George's Math 5 class, she would be functionally illiterate." Similarly, the accidental generalization "All bodies of pure arsenic have a mass of less than one ton" does not support counterfactual statements like "Two bodies of pure arsenic whose combined masses are more than one ton cannot be fused to form one body," or "If two such bodies were fused, the mass of the resulting body would still be less than one ton."

The ontological status of scientfic laws is thus not quite clear. They seem to be more than just summaries of constant conjunctions, since they support counterfactuals, yet they are certainly less than necessary or logical truths.

There is a story by Leo Rosten (1968) that is marginally relevant:

An insensitive oaf (a *bulvon*) was about to go out on a blind date and asked his Lothario of a friend for some advice. The friend responded, "I'll tell you a secret. Jewish girls love three topics of conversation: food, family, and philosophy. That's all you need to remember. To ask about a girl's tastes in food is to make her feel important. To ask about her family shows that your intentions are honorable. And to discuss philosophy with her shows you respect her intelligence."

The *bulvon* was pleased. "Food, family, philosophy!"

He met the girl and blurted, "Hello. Do you like noodles?"

"Why, no" said the startled girl.

"Do you have a brother?"

"No."

The *bulvon* hesitated for just a moment: "Well, if you had a brother, would he like noodles?"

Medical people and biologists sometimes use "cause" in a bizarre way. They reason that if x cures y, then lack of x must cause y. If dopamine, for example, lessens the tremors of Parkinson's disease, then lack of dopamine must cause it. If an antagonist of dopamine reduces the symptoms of schizophrenia, then an excess of it must cause schizophrenia. One is less likely to make this mistake when the situation is more mundane. Few people believe that since aspirin cures headaches, it must be the case that lack of aspirin in the bloodstream causes them.

Similar things can be said about some uses of "cause" in the social sciences.

Two Australian aborigines were brought to this country and saw for the first time a waterskier winding and cavorting his way around a lake. "Why is the boat going so fast?" asked one of the aborigines. The second answered, "Because it is being chased by a madman on a string."

The Tortoise Came First?

In clarifying the structure of the first-cause argument for the existence of (a) God, Bertrand Russell cites the Hindu myth that the world rests on an elephant and the elephant rests on a tortoise. When asked about the tortoise, the Hindu replies, "Let's change the subject."

But let's *not* change the subject. Assuming *some* reasonable understanding of the word "cause," either everything has a cause or something does not. If everything has a cause, God does too. If there is something that does not have a cause, it may as well be the physical world as God or a tortoise.

The cogency of this reply to the first-cause argument is indicated by Saint Augustine's reaction to a version of it. When he was asked what God was doing before He made the world, he answered, "He was creating a hell for people who ask questions like that."

The natural-law argument for the existence of God has a similar structure and is thus open to a similar reply. The argument posits God as the lawgiver, the author of order and law in the natural world. Whatever power the argument has is greatly diminished by asking why God created the particular natural laws that He did. If He did it whimsically, for no reason at all, there is then something that is not subject to natural law; the chain of natural law is broken. On the other hand, if He had reasons for issuing the particular laws that He did, then God himself is subject to law and there is not much point in introducing Him as an intermediary in the first place.

There is a statistical "law of nature" that says that if you throw a pair of dice, you will get a two (one on each die) only about once in thirty-six throws. Many laws of nature are, of course, of this sort. Though they indicate an order or lawfulness on one level, this order or lawfulness exists only because of disorder or randomness on a lower level. Compare the second law of thermodynamics, which states that in any *closed* system, entropy (disorder, roughly), being more probable, is continually increasing. (Living things— little islands of order that replicate themselves, and from which more complex configurations can evolve—are not closed systems, and thus do not constitute a counterexample to the second law of thermodynamics. They eat, are warmed by the sun, etc.)

It is in fact hard, if not impossible, to imagine a *completely* random universe. It seems that in any universe there would necessarily be, on some level, some kind of order or lawfulness, even if only of a very Pickwickian sort. Even in a very messy universe one could describe the mess (assuming one were around, which is not likely) or enunciate some higher-order prediction to the effect that no lower-order predictions seem to work.

Why is it not termed a miracle when a freak gust of wind topples a single flowerpot from a tenth-story window and the flowerpot falls onto the head of a person walking on the sidewalk below, or when a faith healer ministers to a blind man who then becomes lame?

Constants of different sorts: Planck's constant, 55 mph speed limit, 3 ft. in a yard, π.

Laws of different sorts: Conservation of mass-energy, parking ordinance, Thou shalt not kill, Boyle's laws of gases.

One reasonable reaction to the refutation of the first-cause and natural-law arguments is to try to make sense of the first cause not only causing the second cause(s) but also causing itself—or, analogously, the most general law not only explaining the next most general law(s) but also explaining itself. Robert Nozick, in *Philosophical Explanations*, considers one such self-subsumptive principle, P, of the following type : Any lawlike statement having characteristic C is true. Principle P is used to explain why other, less general laws hold true: they hold true because they have characteristic C. What, however, explains why P holds true? The answer is that P, since it also has characteristic C, explains why it itself holds true. In short, P, if true, explains itself.

Even Nozick acknowledges that this "appears quite weird—a feat of legerdemain" (1981). Still, there are not, as he notes, many alternatives. The chain of causes (laws) is either finite or infinite. If it is finite, the most basic cause (most general law) either is a brute, arbitrary fact or is self-subsuming.

Self-subsuming principles need not be any deeper or of any higher order than what they explain. They could perhaps be handled by a variant of Kripke's theory of truth, in which statements attain their level naturally rather than being preassigned a fixed level, and in which self-referential statements sometimes receive truth values. (Perhaps even the oscillation associated with certain paradoxical statements—if they're true, they're false, which makes them true, which makes them false, etc.—could be given a physical meaning consistent with the notion of self-subsumption.)

Nozick writes also of certain yogic mystical exercises that help to bring about the experiential analogue of self-subsumption. He theorizes that "one of the acts the (male) yogis perform, during their experiences of being identical with infinitude, is auto-fellatio, wherein they have an intense and ecstatic experience of self-generation, of the universe and themselves turned back upon itself in a self-creation" (1981).

All this, of course, is contrary to Marxist theory. As Marx (Groucho) himself might have observed, "First Russell defines mathematics as the subject where we never know what we're talking about nor whether what we're saying is true, and now this Nozick joker says that the most basic laws of the universe have something to do with yogis playing with themselves; and yet they call *me* a comedian.''

Pascal once said, "To poke fun at philosophy is to be a philosopher" (1966). Although this is not quite true—poking fun being a necessary but not a sufficient condition for being a philosopher—it and Groucho's imagined reaction to Russell and Nozick do suggest a deep resonance between humor and philosophy. Ideally, both activities require—in fact, presuppose—a free intelligence stepping back from roles, rules, and rote, in order to respond to the world with honesty and courage. Ideally.

Of Birds and Strange Colors

Hume's traditional problem of induction aside (consider it solved, resolved, dissolved, or just ignored), there remains the problem of exactly which regularities in nature are projectable into the future. All samples of water so far examined (under normal atmospheric pressure) have been found to have a freezing point of 32°F; it thus seems reasonable to project this regular connection between being a water sample and having a freezing point of 32°F into the future. It is also true that all major economic depressions have occurred at the same time as large sunspots; yet in this case, it does not seem reasonable to project this regular connection between economic depressions and sunspots into the future.

Nelson Goodman has shown (1965) that the question of which regularities are projectable into the future is more problematical than these two examples indicate. Goodman's projectability paradox can be explained by considering the strange electrical terms "condulator" and "insuductor." A date in the future is selected, say January 1, 2010. Something is termed a condulator if it is a conductor and the time is before 2010, or if it is an insulator and the time is after January 1, 2010. Something is an insuductor, on the other hand, if it's an insulator and the time is before 2010, or if it's a conductor and the time is after January 1, 2010. Now let's consider the properties of copper wire. All sam-

ples of copper wire tested up to now (1984) have been conductors; we therefore feel we have good evidence that all
copper wires are conductors of electricity. But, Goodman
points out, all copper wires so far tested are also condula-
tors: it seems as though we have just as good a body of evidence for the proposition that all copper wires are condu-
lators (and hence insulators beginning in 2010).

Of course, a natural objection to these electrical terms
"condulator" and "insuductor" is that they are defined in
terms of the year 2010. But what if there really were a people who developed a language or scientific theory of which
condulator/insuductor was a part? They could make the
same charge against us. "Conductor," they could maintain,
is an odd term, being defined as condulator before 2010
and insuductor afterward. "Insulator" is just as strange,
being insuductor before 2010 and condulator from then
on.

More generally, it is not just these isolated terms but
the languages, theories, and worldviews of which they're a
part that structure the world and its observers in incommensurable ways, and thus lead to very different expecta-
tions about the future. One could, for example, rig up two
electric chairs, one with copper wiring and the other with
asbestos wiring, and have the leading scientists from each
side sit in the chair of their choice on January 1, 2010.

Goodman's original example concerned the odd color
terms "grue" and "bleen" and is usually referred to as the
grue-bleen paradox. An object is grue if green before 2010,
say, or blue afterward. Bleen is defined similarly, and it's

stressed that whatever evidence we have for all emeralds being green is also evidence for all emeralds being grue. Grue-bleeners, however, can point out that it is really "blue" and "green" that are the strange color terms, the latter being grue before 2010 or bleen afterward, and the former being bleen before 2010 or grue afterward. Of course, an indefinite number of such puzzles can be generated—republicrats and democans (2010 could be a cataclysmic year)—but how to get around them is not completely clear.

Another intriguing paradox is due to Carl Hempel (1965). His "raven" paradox, so called because it is usually illustrated with ravens, can be easily stated. Suppose we want to confirm the statement "All ravens are black." We go out, look for ravens, and check to see if they are black. We believe that if we observe enough instances of black ravens, we will have confirmed (not necessarily conclusively verified) the statement "All ravens are black." But by elementary logic, "All ravens are black" is logically equivalent to "All nonblack objects are nonravens." Since the statements are equivalent, any observation that confirms one confirms the other. But pink flamingos, orange shirts, and chartreuse lampshades are all instances of nonblack objects, and thus tend to confirm the statement "All nonblack objects are nonravens"; therefore they must also confirm "All ravens are black." Hence we arrive at the curious position of having pink flamingos, orange shirts, and chartreuse lampshades confirming the statement that all ravens are black!

What is the problem? Well, it still is not clear to people. Two quick points should be made, however. One is that merely amassing instances of a statement is not enough to confirm it. The second is that nonravens and nonblack objects are much more numerous than ravens and black objects. Perhaps we could understand pink flamingos, orange shirts, and chartreuse lampshades as confirming, but only very slightly, the two equivalent statements above—not as much, in fact as a black raven would.

Similarly, to confirm "All congressmen have problems with grammar" one could go out, look for people who speak grammatically, check to see that they're not congressmen, and obtain some minuscule confirmation for the statement. More conclusive confirmation for it, though, would come from attending congressional hearings.

It is hard to say in general when an observation confirms (again, not necessarily conclusively verifies) a statement. Hempel considered some conditions that the notion of confirmation might be expected to satisfy. Two obvious ones, it seemed, were (1) If an observation o confirms a statement h and h implies another statement k, then o confirms k; and (2) If an observation o confirms a statement k and another statement h implies k, then o confirms h.

But from (1) and (2) we can derive the following: Let h = the theory of relativity, and k = the thermostat is above $80°$. Then its being hot in the room tends to confirm k (on any intuitive understanding of confirmation), and thus tends by (2) to confirm the compound statement (h and k)

(symbolically, $h \wedge k$), since $h \wedge k$ implies k. Hence, by (1) h is confirmed, since $h \wedge k$ implies h. Thus we conclude that its being hot in the room confirms the theory of relativity! Something is obviously wrong with (1) and (2) together. (2) seems especially suspect, but some (weakened) version of (2) is certainly often used, and is necessary in the everyday practice of science. Even (1) is not always the case.

Let us consider one final oddity. Ever since Plato, knowledge has been taken by many philosophers to be justified true belief. A subject S is said to know a proposition P if (a) P is true; (b) S believes that P is true; and (c) S is justified in believing that P is true. Edmund L. Gettier has shown (1963) that these three ancient conditions are not sufficient to ensure knowledge.

To show this, suppose that George and Martha are the only applicants for a certain job at Lower Slobovia State University. Suppose further that George has strong evidence for the following compound proposition:

(1) Martha is the person who will be hired, and Martha has unruly hair this morning. George's evidence for (1) might be that the chairman of the department informed him that Martha's specialty, unlike George's, was the one the department had been looking for. Further, he could see that the department chairman and Martha had established an early rapport despite her unruly hair—while the chairman barely spoke to him, hurried him out of his office, and

muttered something nasty-sounding to his secretary that made her laugh derisively.

Proposition (1) implies:

(2) The person who gets this job has unruly hair. George sees that (1) implies (2) and thus accepts (2) on the basis of (1), for which, as we've seen, he has strong evidence. Clearly George is justified in believing that (2) is true.

So far, so good. But suppose that, unknown to George, he, not Martha, will get the job. (The chairman, imagine, has a strange psychology.) Also unknown to George, his hair was ruffled by the fan in the elevator, causing his cowlick to stand up. Proposition (2) is thus true even though (1) from which it was inferred is false. Now all the following are true: (a) (2) is true; (b) George believes that (2) is true; and (c) George is justified in believing that (2) is true. But, of course, it is quite clear that George does not *know* (2), since (2) is true in virtue of George's own unruly hair, of which he is unaware. Thus justified, true belief does not constitute knowledge.

If justified true belief does not constitute knowledge, the question remains: What does? One answer, due to the philosophers Fred Dretske and Robert Nozick, involves the notion of subjunctive or counterfactual conditionals. They state that S knows P if (a) P is true; (b) S believes P; (c) If P weren't true, S would not believe P; and (d) If P were true (but other minor things were different), S would still believe P. For example, by these criteria George does not know "The person who gets the job has unruly hair." If the proposition in question weren't true—say, George's hair was not accidently ruffled by the fan in the elevator but he still got the job—George would still believe the proposition. Thus condition (c) is not met, and George cannot be said to know the proposition. (Condition [d] handles other related difficulties.)

Since these four conditions, it can be argued, are sufficient for knowledge, and since such counterfactual and subjunctive conditionals are also helpful in the analysis of scientific law (and notions like "possible worlds"), why not declare the war over and decree that these problems are resolved? The reason, for many philosophers, is that the analysis of these conditionals is at least as problematic as the notions they are meant to clarify.

Truths, Half-Truths, and Statistics

Benjamin Disraeli coined the phrase "Lies, damn lies, and statistics," and the phrase (as well as the sentiment) has lasted—though I like "Truths, half-truths, and statistics" better. In any case, even relatively simple applications of statistics can cause problems, not to mention the horrors associated with things like the often misinterpreted SPSS computer software (Statistical Programs for the Social Sciences).

Probability and statistics, like geometry and mathematics in general, come in two flavors: pure and applied. Pure probability theory is a formal calculus whose primitive terms are uninterpreted and whose axioms are neither true nor false. These axioms originally arise from and are made meaningful by real-life interpretations of terms like "probability," "event," and "random sample." The problem with applying probability and statistics is often not in the formal mathematical manipulations themselves, but in the appropriateness of the application, the validity of the interpretation, and indeed the "reasonableness" of the whole enterprise. This latter activity goes beyond mathematics into the sometimes murky realm of common sense and the philosophy of science (grue-bleen, ravens, etc.). Even though 1 plus 1 equals 2, one glass of water plus one glass of popcorn does not equal two glasses of mixture. The mathematics is fine, the application is not.

MARTHA: What did you get for the density of the block, George?

GEORGE: Well, it weighed about 17 pounds and had a volume of about 29 cubic feet, so I guess the density is .58620689551 pounds per cubic foot. This calculator's really swell.

Babe Ruth and Lou Gehrig played baseball for the New York Yankees. Suppose Ruth had a higher batting average than Gehrig for the first half of the season. Suppose further that during the second half of the season Ruth continued to hit for a higher batting average than Gehrig. Is it nevertheless possible for Gehrig's batting average for the entire season to be higher than Ruth's batting average for the entire season? The fact that I've used up a paragraph asking the question indicates that the answer is yes, but how can it be?

One way it can be is for Ruth during the first half of the season to hit for an average of .344, getting 55 hits in 160 times at bat; while Gehrig during this same time hits for an average of .342, getting 82 hits for 240 times at bat. During the second half of the season Ruth's average is .250, since he gets 60 hits in 240 times at bat; whereas Gehrig's is

.238, as he gets 38 hits in 160 times at bat. Nevertheless for the season as a whole, Gehrig's average of .300 is higher than Ruth's average of .287.

Thus even the third-grade notion of an average can be misused, not to mention (as I already have) things like complicated multidimensional analyses of variance.

If Waldo comes from country x, 30 percent of whose citizens have a certain characteristic, then if we know nothing else about Waldo, it seems reasonable to assume that there is a 30 percent probability that Waldo shares this characteristic. If we later discover that Waldo belongs to a certain ethnic group 80 percent of whose members in the region comprising countries x, y, and z have the characteristic in question, what now are Waldo's chances of sharing this characteristic? What if we subsequently determine that Waldo belongs to a nation-x-wide organization only 15 percent of whose members have this characteristic? What now can we conclude, with all this information, about Waldo's chances of having the characteristic in question?

Wittgenstein writes about a man who, not being certain of an item he reads in the newspaper, buys a hundred copies of the paper to reassure himself of its truth. Given the extent to which news-people and the media report each other's reports, checking in several different papers or periodicals is not much more intelligent.

News release: Abortions are becoming so popular in some countries that the waiting time to get one is lengthening rapidly. Experts predict that at this rate there will soon be a one-year wait to get an abortion.

The projection of "trends" linearly into the future is often about as reliable as this "news release."

Most automobile accidents occur close to home, so we can see that near one's home is the most dangerous place to drive.

Very few accidents occur when one is driving over 95 mph, so it's clear that driving this fast is actually quite safe.

It is a surprising, almost counterintuitive, fact that if just 23 people are chosen at random from a telephone directory, the probability is .5 (chances are about 50-50) that at least two of them will have the same birthday. Recently someone was trying to explain this oddity on a television talk show. The incredulous host thought the man must be wrong, and so asked the studio audience how many people had the same birthday as he did, say March 19. When nobody in the audience of about 150 people responded, the host felt vindicated and the guest felt embarrassed. Actually, the question the host had raised was very different fromn the one the guest had been discussing. It turns out that a randomly selected group of 253 people is required in order for the probability to be .5 that at least one member of the group has any *specific* birthdate (such as March 19); a group of only 23 people is required for the probability to be .5 that there is *some* birthdate in common.

This is a specific instance of a very general phenomenon. Even though any particular event of a certain sort may be quite rare, that *some* event of that sort will occur is not rare at all. The science writer Martin Gardner illustrates this point with the story of a spinner that is equally likely to stop at any of the twenty-six letters of the alphabet. If the spinner is twirled 100 times and the results recorded, the probability of any *particular* three-letter word, say "cat," appearing is quite small, whereas the probability of *some* three-letter word appearing is very high.

Columbus discovered the New World in 1492, while his fellow Italian Enrico Fermi discovered the atomic world in 1942. John Kennedy, elected president in 1960 and assassinated in office, had a Lincoln for a secretary, while Abraham Lincoln, elected president in 1860 and assassinated in office, had a Kennedy for a secretary. As Gardner has noted, the acronym formed by the planets listed in order—Mercury, Venus, Earth, Mars, Jupiter, Saturn, Uranus, Neptune, Pluto—is M V E M J <u>S U N</u> P, while that for the months is J F M A M J J A S O N D. In each case we have an unlikely happening whose type, while almost impossible to specify precisely, is quite likely to have some instances. This has some relevance to evolution. That a particular branch should have evolved in a particular way is quite improbable. That some branch should have evolved in something like the way in question is much less so.

The probability of getting at least one head on two flips of a coin is .75. The chances of rain tomorrow are 75 percent. I think the odds of George's marrying Martha are 3 to 1. Does "probability = .75" mean the same thing in each of the above cases? The preceding discussion concerning unlikely coincidences might be summarized aphoristically as, "It's very improbable that no improbable event will occur." Are both uses of "improbable" in this statement the same?

Duhem, Poincaré, and the Poconos-Catskill Diet

Even in the best of circumstances, applying the classical canons of scientific inference is not always straightforward. If a certain hypothesis H implies or makes more likely some event I, then if that event I occurs, H is strengthened; whereas if the event I does not occur, H is weakened or refuted. The raven and confirmation paradoxes demonstrated that I's occurrence (the sighting of pink flamingoes, say) does not necessarily strengthen H (the statement that all ravens are black). The French philosopher Pierre Duhem, on the other hand, showed that I's *non*occurrence does not necessarily refute or weaken H either.

To see this, consider the Poconos-Catskill diet. Dr. Poconos, a Greek doctor from Pennsylvania, and Dr. Catskill, an Irish doctor from New York, prescribe two large servings of pastitsio (a meat, cheese, and noodle dish), three pieces of baklava, and four beers for every meal. They guarantee that after one week on this diet, a person will lose at least six pounds. George and Martha go on this diet and gain nine pounds in one week. Must Drs. Poconos and Catskill retract their hypothesis H, that the diet is effective, given that I, the loss of at least six pounds, did not occur? Of course not. They could maintain that a host of other auxiliary, but tacit, hypotheses failed, and not their pet hypothesis H, the Poconos-Catskill diet. Maybe the pastitsio was too heavily salted or not salted enough, maybe George and Martha slept fourteen hours a day during the week, or maybe the meals were not spaced correctly.

Thus the nonoccurrence of I can never by itself refute H, since there will always be scapegoats—auxiliary hypotheses to blame. What is being tested is never just "H implies I," but rather "H and H_I and H_2 and H_3 and . . . implies I," where the ellipsis indicates indefinitely many auxiliary hypotheses. I's nonoccurrence thus indicates only that either H or (at least) one of the auxiliary hypotheses must be false, not necessarily that H is false.

Willard Van Orman Quine has gone even further than Duhem in maintaining that experience never forces the rejection of individual statements. He conceives of science as an integrated web of statements, procedures, and formalisms in contact with reality only at its edges. Any impact of the world on the web is distributed throughout the web, with no part (even logic) being absolutely immune and no part having to bear the brunt of that impact alone. Adjustments can always be made in the whole web to accommodate the experience, but there is no unique way to make these adjustments—simplicity, efficiency, and tradition being some of the criteria for a good web (science). We could thus accept that the Poconos-Catskill diet is an effective weight-loss program (as commercials often phrase it), but we would have to make fairly drastic changes in the rest of our web.

Similarly, in physics, if one wanted, for whatever reason, to accept some bizarre statement as true, one could do so by suitably and radically altering other statements, the interpretation of certain physical terms, and so on. This is related to the traditional problem in the philosophy of science of where to draw the line between empirical physics and a priori geometry. If one insists that geometry must be Euclidean, one's physics (in certain astronomical contexts, say) becomes quite strange, with accelerations and forces that don't make sense by traditional theoretical standards. An alternative is to switch to an appropriate non-Euclidean geometry that makes the physics simpler, but which itself seems quite counterintuitive at first. Which geometry-physics combination to use depends on one's purposes and is to some extent a matter of convention, as Henri Poincaré first noted (1913).

Of course, not just physics but any complicated phenomenon, especially one that involves human interaction, admits of many different and incompatible interpretations, each consistent with reality. The differing "theories" of their particular marriage entertained by a husband and wife, for example, often manifest this incomparability, as does the following Sufi story adapted from Masud Farzan's *Another Way of Laughter*:

Since a Roman scholar was visiting Timur's court, the emperor asked the Mulla to ready himself for a battle of wits with the scholar.

The Mulla piled books with fictitious but impressive-sounding titles on his donkey, books such as *The Theory of Universal Bifurcants*, *Erosion and Civilization*, *A Critique of Tolerant Purity*, and *Social Origins of Mental De-activation*.

On the day of the contest, the Mulla appeared in court with his donkey and books. His native wit and intelligence overwhelmed the Roman scholar, who finally decided to test the Mulla's knowledge of theoretical matters. The Roman scholar held up one finger.

The Mulla answered with two fingers.

The Roman held up three fingers.

The Mulla responded with four.

The scholar showed his whole palm, to which Mulla responded with a closed fist.

The scholar then opened his briefcase and took out an egg. The Mulla responded by digging an onion out of his pocket.

The Roman said, "What's your evidence?

The Mulla answered, "*The Theory of Universal Bifurcants*, *A Critique of Tolerant Purity*," etc.

When the Roman sputtered that he'd never even heard of those titles, the Mulla responded: "Of course you haven't. Look and you will see hundreds of books you've never read."

The Roman looked and was so impressed that he conceded defeat. Since no one had understood any of this, later, after refreshments had been served, the emperor leaned over and asked the Roman scholar the meaning of it all.

"He is a brilliant man, this Mulla," the Roman explained. "When I held up one finger, meaning that there was only one God, he held up two to say that He created heaven and earth. I held up three fingers, meaning the conception-life-death cycle of man, to which the Mulla responded by showing four fingers, indicating that the body is composed of four elements—earth, air, water, and fire."

"Well then, what about the egg and the onion?" the emperor pressed.

"The egg was the symbol of the earth (the yolk) surrounded by the heavens. The Mulla produced an onion, indicating the layers of heavens about the earth. I asked him to support this claim to assign the same number of layers of heavens as there are layers of onion skin, and he supported his claim by all those learned books of which I alas am ignorant. Your Mulla is a very learned man indeed." The dejected Roman then departed.

The emperor next asked the Mulla about the debate. He replied, "It was easy, Your Majesty. When he lifted a finger of defiance to me, I held up two, meaning I'd poke both his eyes out. When he held up three fingers indicating, I'm sure, that he'd deliver three kicks, I returned his threat by threatening four kicks. His whole palm, of course, meant a slap in my face, to which I responded with my clenched fist. Seeing I was serious, he began to be friendly and offered me an egg, so I offered him my onion."

Whether Mulla or Roman scholar was "right" is a matter of convention or tradition. As noted, conventionalism is the view that scientific laws are to a significant extent disguised conventions that reflect a decision, for one reason or other, to adopt one of various possible descriptions of a phenomenon. To return to physics, in some cases to ask if one event precedes another makes as much sense as asking if New York is really to the right of Chicago.

George Carlin once listed six reasons for doing something or other: I, b, III, four, E, vi. Notation, though clearly a matter of convention, is crucial. Imagine trying to explain even some elementary mathematics, such as the quadratic formula, without good notation.

If the story of the Mulla and the Roman seems too farfetched, imagine a modern physicist trying to explain to an Australian aborigine that neutrinos have no mass, or something about the properties of quarks and black holes.

Reductionism, Fallibilism, and Opportunism

There are, of course, in addition to conventionalism many other "isms" in the philosophy of science, some of which should at least be mentioned. The most common of these is opportunism, as illustrated by the following well-known story:

MARTHA: What are you doing out by the streetlight, George?

GEORGE: I'm looking for my car keys.

MARTHA: But you dropped them on the grass near the bushes.

GEORGE: I know, but the light is better out here.

The following story due to Leo Rosten (1968) makes the same point, and also says something about the organization and contents of this book:

A famous rabbi was asked by an admiring student "How is it you always have a perfect parable for any topic?"

The rabbi smiled and said "I'll answer with a parable," and told this story:

Once there was a lieutenant in the tsar's army who, riding through a small shtetl, noticed a hundred chalked circles on the side of a barn, each one with a bullet hole in the center. The astonished lieutenant stopped the first passer-by and inquired about all the bullseyes.

The passerby sighed. "Oh that's Shepsel, the shoemaker's son. He's a little peculiar."

"I don't care. Anyone who's that good a shot . . ."

"You don't understand," interrupted the passerby. "You see, first Shepsel shoots and then he draws the chalk circle."

The rabbi smiled. "That's the way it is with me. I don't look for a parable to fit the subject. I introduce only subjects for which I have parables."

Behaviorism is the philosophy of social science doctrine to the effect that psychological functioning is definable in terms of overt behavioral manifestations. It, together with a mindless respect for statistical formalism, has led to the publishing of whole barrels full of trivial research of the following form: If property x (humor, say) is operationally defined in this way (number of chuckles elicited by a book of cartoons), and property y (self-sufficiency, say) is operationally defined in this way (number of yes answers on some "self-sufficiency questionnaire"), then the correlation coefficient between x and y is .621 (at least for students in Prof. George's 8:30 psychology class).

The existence of boredom implies the falsity of behaviorism, but the details of the derivation bore me.

Behaviorism is a type of reductionism, the latter being any doctrine that claims to reduce what is apparently more sophisticated and complex to what is less so. Sometimes, of course, this is possible; large parts of genetics can be reduced to molecular biology, and some parts of thermodynamics to statistical mechanics, for example. Sometimes this reduction is not possible, at least not in any natural manner.

Consider, for example, clocks or can openers. (I wonder if the preceding seven-word sentence has ever appeared in print before?) A clock is any device for keeping time, a can opener any device for opening cans. To characterize these objects in purely physical terms, to reduce all talk of clocks and can openers to the language of physics, would be a pointless (though theoretically possible) endeavor. Imagine what sort of physical description could encompass all and only sundials, grandfather clocks, wristwatches, digital clocks, etc. Any such purely physical description would be a messy, unsystematic hodgepodge of ad hoc physical statements, and thus would offer no illumination on the purposes or identification of clocks or can openers.

Conversely, just as it is pointless though theoretically possible to describe clocks and can openers in purely physical terms, it is pointless though theoretically possible to eliminate theoretical physical terms like "neutrino," "guard," and "covalent bond" and replace them with an unsystematic hodgepodge of ordinary, everyday observational statements containing terms like "red," "cold," and "hard." The predictive and organizational strength of scientific theories would, of course, be lost if this were done. Carl Hempel (1963) has shown explicitly how to rid science of theoretical terms (and thus cripple it) using Craig's theorem, a well-known result from mathematical logic. For some reason, though, one never hears this referred to as reducing physics to the "theory of common sense."

Richard Dawkins in his book *The Selfish Gene* seems to argue that our genes try to perpetuate themselves by fashioning our actions in such a way as to assure their own survival. Human behavior and culture is, on this sociobiological view, largely determined by our genes' desire for self-preservation. This reductionist account certainly needs supplementation. Note, for example, that different cultures with essentially the same genetic pool engender quite different behaviors. Note, too, that we can carry the reduction even further. One might argue that it is not our genes that are selfish, but a particular chemical bond in the genes. Our actions are not determined by our genes after all, but by a chemical bond in them that is trying to perpetuate itself. I am being simplistic, but so is the thesis as it is usually presented.

x can be explained in terms of *y*. *y* does not have property *P*. So *x* does not have *P*.

The greenness of grass, the blueness of the sky, the flesh tones of a human face can all be explained in terms of (the properties of) atoms—frequencies and so on. But atoms themselves are not colored. So grass is not green, the sky is not blue, and faces are not flesh-colored.

Similar arguments have been used to show that values, ethics, ideals, even intentions and beliefs, are illusions.

There is a story about a London man who spent his life looking around and recording everything he noticed in a series of notebooks. He directed that after his death these observation notebooks be forwarded to the Royal Society, so that scientists there could use them to fashion a new scientific theory.

Though some people still think that science advances in this way (consider some behaviorist research, e.g.), most realize that scientific research needs first a focus or problem; next, assumptions and hypotheses; and only then, new observations. Scientific pronouncements should be supported (or at least supportable), and, as the English philosopher Sir Karl Popper has stressed, they must be capable of being shown false. They must be falsifiable or fallible, at least in principle.

When a plane crashes, many people remark that these sorts of things always happen in threes. This belief is, as stated, completely unfalsifiable. Even if the next crash, two months later, occurs in Peru, and the third involves a small private plane in an Arkansas cornfield, these people can still maintain that they have some special insight into aeronautical malfunctioning, or, more commonly, that "everyone" has this knowledge. Similar things can be said about remarks like "Whatever God wills, happens."

Popper has criticized Marxism and Freudianism for being too much like the above theory of plane crashes—that is, for not being falsifiable and thus not being real sciences. He has in mind moves of the following sort: A Marxist predicts that the "ruling class" will respond in such and such an exploitive way to some crisis; when it doesn't, this is attributed to some sneaky, co-opting policy of the ruling class. Similarly, an orthodox psychoanalyst might predict a certain kind of neurotic behavior; when his patient doesn't behave in this way, but in a quite contrary way, this is attributed to a "reaction-formation." Popper does not quite say so, but he hints that Freud is a fraud and that Karl Marx makes less sense than Groucho.

More generally, Popper is opposed to historicism, the doctrine that there are immutable "laws of development" that govern the historical process and allow long-term social forecasts. One argument for Popper's position is that scientific advances that obviously greatly affect social development are not predictable. How could someone predict Descartes's discovery (or invention, if you like) of analytic geometry, or Einstein's discovery-invention of relativity theory? Imagine a "futurologist" of the sixteenth (or nineteenth) century predicting the development of analytic geometry (or relativity theory): if such a prediction had any detailed content, it would in a sense already be fulfilled and thus not be a prediction at all.

Less well known than Popper's ideas on falsifiability is his concern with the notion of randomness, the topic of the next section.

Randomness and the Berry Task

GEORGE: What's Waldo's phone number? I always forget.

MARTHA: Let me see. The Goodmans have 2 children, the Frankels have 3, the Passows have 7 dogs (though Myrtle may be pregnant), the Youngs have 2 kids from his first marriage, 1 from hers, and 3 of their own, and the Sturms have 9 pets in all. Waldo's number is 237-2139.

GEORGE: Thanks. It's a good thing I know the area code.

(1) 00100100100100100100100100100100100100100 . . .

(2) 10110101011011011010101101010101101011010 . . .

(3) 10001011011011000101011001011110100101110 10 . . .

Why is it that the first sequence of 0s and 1s is likely to be termed orderly or patterned, the last sequence random or patternless, and the second somewhere in between? Trying to answer this simple question will lead to a definition of randomness, an important insight in the philosophy of science, and an alternative proof of Godel's incompleteness theorem.

Returning to the question, note that the first sequence has an easily expressed pattern: two 0s, then a 1, repeated indefinitely. The third sequence has no such pattern; while in the second sequence, 0s always alternate with either one or two 1s, but the occurrence of one or two seems pattern-less.

With examples like this in mind, the American computer scientist Gregory Chaitin and the Russian mathematician A. N. Kolmogorov defined the *complexity* of a sequence of 0s and 1s to be the length of the *shortest* computer program that will generate the sequence in question. For the sake of uniformity, assume that the language of the program is coded into a sequence of 0s and 1s so that the program itself can be considered a sequence of 0s and 1s. Thus we can take the length of a program to be simply the number of 0s and 1s in it. If the program is the shortest one to generate a given sequence, the program's length in bits (0s or 1s) is the complexity of that given sequence.

A program that generates the first sequence above will just be a translation into machine language of the following recipe: two 0s, then a 1, repeated x times. This program should be quite short compared to the length of the sequence, which, let us assume, is 10 trillion bits; and thus the sequence has, despite its length, a complexity of only 1 million bits, say.

A program that generates the second sequence would be a translation of the following: one 0 alternating with either one or two 1s, the pattern of 1s being one-two-one-one-two-two-two-one-one-two-one-one-one-two-one-two-one-one . . . If the second sequence were very long, say 10 trillion bits again, and this pattern continued, any program that generated it would have to be quite long to give the pattern of the intervening 1s. Still, the shortest such program, considered as a sequence of 0s and 1s itself, would be considerably shorter than the 10-trillion-bit sequence it was generating. The complexity of this sequence thus might be only 5 trillion, for example.

With the third sequence the situation is different. The sequence, let us assume, is so disorderly throughout its 10-trillion-bit length that no program we might use to generate it would be any shorter than the sequence itself. Since all the program can do in this case is just list the bits in the sequence, there is no way it can be shortened. Such a program, itself expressed in bits, is at least as long as the sequence it is supposed to generate, whose complexity thus is at least 10 trillion. The sequence is random.

More formally, we can define a sequence to be *random* if its complexity is (roughly) equal to its length; that is, if the shortest program capable of generating it has (roughly) the same length as the sequence itself. A sequence is then not random if its complexity is less than its length. Sequence (3) is random on this account, while sequences (1) and (2) are not. Two important consequences of these definitions are (a) If two sequences of different lengths are random, the longer one is more complex; and (b) For any given whole number x, the vast majority of sequences of bits of length x are random; there are only a relatively few low-complexity ones of any given length.

Intelligibility or precision: to combine the two is impossible.
Bertrand Russell

Attempts have been made to apply these formal notions of complexity and randomness more generally. A. Solomonoff theorized, for example, that a scientist's observations could be coded up into a sequence of 0s and 1s. The goal of science then would be to find short programs (algorithms, recipes) capable of generating (deriving, predicting) these observations. Such a program, so the story goes, would be a scientific theory, and the shorter it was, relative to the phenomena it predicted, the more powerful it would be. Random events would not be predictable, except in a very fractured sense by a program that simply listed them.

There are, of course, serious problems in trying to extend the use of these technical notions to more general contexts. Where do these sequences of 0s and 1s come from in the first place? Exactly how are observations to be coded into a sequence of bits? Or prediction sequences decoded? What relation do they have to other significant categories? (Recall the sportscaster in a hurry: "And now for the baseball scores—6 to 2, 4 to 1, 8 to 5, 7 to 3, 5 to 0, and in a real slug-fest, 14 to 12.") It is impossible to divorce these sequences from the way they are obtained and employed, and from the human interests and values that led to their discovery and explanatory significance. Without this supporting scientific and cultural background, the sequences are meaningless.

Wittgenstein once remarked, "That Newtonian mechanics *can* be used to describe the world tells nothing about the world. But this does tell us something—that it can be used to describe the world *in the way in which we do in fact use it*." The same thing can be said about scientific theories conceived of as programs that generate predictions: the way in which we do in fact use them *is* their scientific content. The reductionist tendency to be seduced by the simplicity and exactitude of this account of scientific theories as pre-diction-generating programs, and the consequent desire to declare that science is nothing but the study of such pro-grams, should be resisted.

Ah but a man's reach should exceed his grasp,
Or what's a heaven for.   —*Robert Browning*

The formal notions of complexity and randomness, though they suffer from the limitations discussed above, are nevertheless often suggestive and useful. Chaitin has employed them and a detoxified version of the Berry para-dox to give an alternative proof of Godel's first incomplete-ness theorem. Since the proof sheds a somewhat different light on this famous result, a sketch of it follows.

The Berry paradox, first published in 1908 by Bertrand Russell and attributed to a Mr. Berry, asks one to consider the following task: "Find the smallest whole number that requires in order to be specified more words than there are in this sentence." Examples such as the number of hairs on my head, the number of different states of a Rubik cube, and the speed of light in centimeters per decade, each specify, using fewer than the number of words in the given sentence, some particular whole number. The paradoxical nature of the task becomes clear when we realize that the Berry sentence specifies a particular whole number that, by definition, it contains too few words to specify.

What yields a paradox in English can be modified (detoxified) to yield a statement in a formal system that cannot be proved, and whose negation also cannot be proved. Consider a formal axiomatic system of arithmetic expressed in some formal language that contains symbols for addition, multiplication, and so on. This system—formulas, axioms of arithmetic, rules of inference—can be encoded into a sequence of Os and Is, a binary program P whose length in bits is $L(P)$. We can then conceive of a computer executing this program and over time generating from it the theorems of the system (encoded, of course, in bits). Stated a little differently, the program P generates sequences of bits that we interpret as the translations of statements in arithmetic, statements that the formal system has proven.

Now we ask whether the system is complete. For every arithmetic statement A, is it the case that either A or its negation, $\sim A$, will always be a theorem? (Is the sequence of bits associated with either A or $\sim A$ eventually generated by the computer?)

To see that the answer to this question is "No," we make a crucial alteration of the Berry sentence: we replace "that requires," a metalevel notion expressed in English, by "that can be proved to require," an object-level notion expressible in terms of 0s and 1s. (We assume that enough arithmetic is included in the original formal system to permit "talk," via some reasonable code involving 0s and 1s, of notions like provability, complexity, etc. within the system.)

Recalling that the complexity of a sequence of bits is the length of the shortest program needed to generate the sequence, we find that even the altered Berry task is impossible, though not paradox-inducing: "Find—that is, generate—a sequence of bits that can be proved to be of complexity greater than the number of bits in this program." (Again, enough arithmetic is assumed to be included in the original formal system to allow, via some sort of coding, for the self-reference of "in this program.") The program cannot generate such a sequence, since any sequence that the arithmetic program P generates must, by definition of complexity, be of complexity less than P itself is. Stated alternatively, there is a limit to the complexity of the sequences (translations of arithmetic statements) generated (proved) by P. That limit is, by the definition of complexity, the length, $L(P')$, of the shortest program encoding P.

Since a sequence is random if the length of any program generating it is at least as long as the sequence itself, we can further conclude that a formal system can generate a random sequence only if the sequence is less complex than the sequence encoding the system. That is, the only random sequences that the arithmetic system P can generate are necessarily less complex than P.

Hence we finally get to the undecidable statements that Godel promises us. Take g to be a random sequence of bits of complexity greater than that of P encoded into bits. This is always possible, since for any whole number x, most sequences of length x are random. Then the statement G: "g is random," properly encoded into bits, is unprovable by the system P. (Remember, "random" is defined in terms of "complexity," which is defined in terms of lengths of programs, all of which talk can be translated into arithmetic and then via code into a sequence of bits. Thus the statement "g is random" can be translated into a sequence of bits.) P cannot generate the sequence that is the translation of "g is random" for the reason discussed above; the sequence is too complex. Neither can P generate the sequence that represents the negation of "g is random," because g is random; since the axioms of P are true and the rules of inference preserve truth, only true statements can be proved. Therefore G can be neither proved nor disproved.

Godel's theorem can thus be interpreted as a consequence of the limited complexity of any formal arithmetic system, a limitation affecting human minds as well as machine programs. There is not the temptation, on this interpretation of the theorem, to dismiss many of the goals of artificial intelligence as impossible in principle, since machines cannot "step outside themselves" to the metalevel. This latter attempted refutation of mind-as-machine is rendered superficially plausible by more-standard proofs of Godel's theorem. Lastly, Chaitin's proof suggests that progress in mathematics is at times not so different from progress in other sciences. Instead of looking for new facts, one looks for new, true, independent axioms whose addition will make the relevant formal systems (or rather their translations into bits) more complex.

Determinism and
Smart Computers

Although falsifiability and verifiability (testability for short) are important properties, it is unwise to be too rigid in interpreting these terms or too hasty in rejecting any pronouncements that are not testable (under any interpretation). Metaphysics is untestable and, as Friedrich Waissman responded in countering some zealous logical positivists, "To believe that metaphysics is nonsense is nonsense."

The metaphysical doctrines of determinism and indeterminism are certainly not testable in any strong sense. How is one to test that every event is determined, or that at least one event is not, *whatever* is meant by "determined"? Certain stronger, more precise theses *can* be formulated and refuted, though. Consider, for example, provable determinism, which maintains that every question in an appropriate language of science can be decided one way or the other on the basis of certain physical and mathematical laws. This is simply false by Godel's theorem, since incompleteness infects even formalized physical theories that include arithmetic in their formalization. A doctrinaire determinist can still maintain that the answer to every question is "determined" by "states of affairs," but not that every answer can be proved to follow from the given physical laws and mathematical theories.

Consider a computer, the "IBM-Cyber-007-know-it-all-smarty-pants," into which has been programmed (in some suitable language) the most complete scientific knowledge of the day, the initial conditions of all particles, and elaborate mathematical techniques and formulas. Imagine further that "know-it-all" answers only Yes or No questions, and that its output device is constructed in such a way that a Yes answer turns off an attached lightbulb and a No answer turns it on. If one asks this impressive machine something about the external world, the machine will respond, let us assume, flawlessly. However, if one asks it if its lightbulb will be on in one hour, "know-it-all" is stumped and cannot answer either way. This question, at least, is "undetermined" by the laws and axioms of its program (although an onlooking computer might be able to answer it).

Related to the "know-it-all" computer is the following phenomenon: In predicting what a person will decide, it is often very important to keep this predictive "information" secret from the person deciding. The scare quotes around "information" are meant to indicate that this peculiar type of information loses its value, becomes obsolete, if given to the person involved; it changes the person. The information, while it may be correct and true, is not universal. The onlooker and the deciding agent (compare "know-it-all") have complementary viewpoints. As D. M. MacKay has written, "To us, our choice is logically indeterminate until we make it. For us, choosing is not something to be observed or predicted, but to be done."

It is suggestive that in quantum mechanics, where indeterminism and subject-object interaction also play a crucial role, we also have the phenomenon of complementary viewpoints. A scientist might know the position of a particle, but that knowledge is incompatible with knowledge of the particle's momentum since, as the Heisenberg uncertainty principle states, the process of determining one necessarily affects the other.

Statistical laws have been around for a long time and may reflect either our ignorance of things or the nature of things. In philosophical parlance, they may reflect either an epistemic or a metaphysical condition. As far as microphysical phenomena go, quantum mechanics points to the latter conclusion. There is a fundamentally and irreducibly probabilistic aspect to atomic and subatomic happenings; some phenomena simply are random. Whether this microphysical indeterminism somehow percolates "up" through the structured complexity of the human (or inhuman) brain, resulting in free will, is an open question.

In any case, just as the mathematical existence of long random sequences of bits was used in the last section to establish Godel's theorem, and thus the falsity of provable determinism, the physical existence of random subatomic happenings demonstrates the untenability of determinism proper.

Some have been unwilling to accept this and have objected, as Einstein did, that "God does not throw dice," that there must be some "hidden variables" whose values, were we to know them, would lead to completely deterministic predictions. That this is logically and physically impossible is demonstrated by self-implicating questions and observations and the quantum mechanical facts. Just how extraordinary one such fact is, is clarified by an inequality due to J. S. Bell.

Bell's Inequality and Weirdness

> "If any of them can explain it," said Alice, "I'll give him six-pence. I don't believe there's an atom of meaning in it."
>
> "If there's no meaning in it," said the King, "that saves a world of trouble, you know, as we needn't try to find any."

This section will demonstrate, I hope, that one of the least problematical aspects of quantum mechanics is its probabilistic nature. More serious is the fact that it almost makes no sense. Almost. Imagine a device having three unconnected parts, as in figures 1 and 2 below. The part C in the middle has a button that, when pressed, sends out a particle to each of the detecting devices A and B. The dials on A and B can be set before the button is pressed, or even while the particles are in flight. (The distance between A and C need not equal the distance between B and C, and the dial for B can even be set after a particle arrives at A but before one arrives at B.) When the particles do arrive at A and B, a light flashes. The light, which may be red or green, flashes no matter how the dials on A and B are set, though whether it is the red or green light that flashes may depend on the dial settings. The dials are set independently of one another.

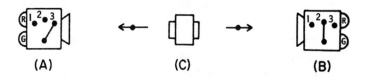

Figure 1. The complete device. *A* and *B* are the two detectors. *C* is the box from which the two particles emerge.

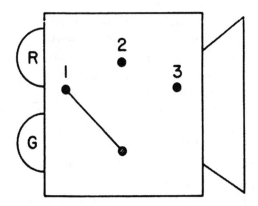

Figure 2. A detector. Particles enter on the right. The red (*R*) and green (*G*) lights are on the left. The switch is set to 1.

The nature of the particles, the construction of the device, in fact all the technical details are irrelevant. All that is important is that such a device can be built with absolutely no connections between the three parts *A*, *B*, and *C*. When the button is pressed, the particles are ejected and the colors of the lights on *A* and *B* are recorded. If this is repeated many times we can discern some pattern to the flashings. We indicate the outcome of any single episode by using a simple notation (due to N. David Mermin [1981], whose general approach I am following): 21*RG*, for example, means that *A*, whose dial was set at 2, flashed *R* while *B*, whose dial was set at 1, flashed *G*. The record of many repetitions thus would look like the following: 32*RG*, 21*GR*, 33*RR*, 22*RR*, 13*RG*, 32*GR*, 23*GG*, 12*RG*, 12*GG*, 11*GG*, 21*GR*, 22*GG* . . .

(An analogy might be helpful. Imagine two psychologists *A* and *B* in different cities, say Albuquerque and Buffalo. Married couples in Chicago separate and travel to these cities, one spouse to Albuquerque, the other to Buffalo, in order to be asked exactly one of three Yes or No questions. The psychologists' records after examining many such couples would look like the following: 31*YY*, 21*YN*, 22*NN*, 13*NY*, 11*NN*, 21*YY*, 21*NY*, 33*YY*, 21*NY* . . .

21*YN* would mean, as in the particle case, that the psychologist in Albuquerque asked question 2 and received a Yes response while the psychologist in Buffalo asked question 1 and received a No response.)

On examining this record we would notice several things: (a) When the dials on *A* and *B* have the same setting (11, 22, 33), the detectors *always* flash lights of the same color, either both red or both green. (b) For these settings, red and green pairs of flashes appear randomly with equal frequencies. (c) When the dials on *A* and *B* have different settings (12, 13, 21, 23, 31, 32), we note that the detectors flash lights of the same color only one-fourth of the time, red and green pairs of flashes appearing randomly with equal frequencies. (d) Three-quarters of the time that the dials are set differently, the lights that flash are of different colors, *RG* and *GR* appearing randomly with equal frequencies.

It is important to remember observation (c), that when the dials on *A* and *B* are set differently (12, 13, 21, 23, 31, 32), these detectors flash the same color only one-fourth of the time. Millions of trials can be run to be certain of this proportion.

Given all this, so what? Well, some strange conclusions seem forced on us. We conjecture that properties of each particle determine the color its detector will flash for each of the three dial settings. This seems to be the only way to explain why the detectors *always* flash the same color when their settings are the same: the detectors *A* and *B* must be responding to some (shared) property of the two particles—size, speed, spin, whatever.

(If, unlikely as it may seem, the members of each and every married couple were always to answer the same way, either both Yes or both No, whenever the psychologists in Albuquerque and Buffalo asked the same-numbered question, it would seem natural to conclude that the psychologists were measuring some real property of these couples.)

It would thus appear that there are eight types of particles—*RRR, RRG, RGR, GRR, GGR, GRG, RGG, GGG*—and that when the dial settings are the same, the flashing of the same color on both detectors indicates the same property present for both particles. Thus if *RRG* particles are ejected by *C* and if the dials of both detectors are set at 2, the red light will flash at both *A* and *B*. If the dials are both set (independently, remember) at 3, green lights will flash at both. On the other hand, if *A*'s dial is set at 1 or 2 and *B*'s at 3, *A* will flash red and *B* green. Similarly, if *GGG* particles are ejected, both detectors will flash green no matter what their settings.

(To continue with the married couples analogy, we conclude that there are eight types of couples—*YYY, YYN, YNY, NYY, YNN, NYN, NNY, NNN*. Consider a couple of type *YNY*. If the members of this couple leave Chicago, arrive at Albuquerque and Buffalo, and are both asked question 3, they will both answer Yes. If only the Albuquerque individual is asked question 3, while the Buffalo member is asked question 2, the latter will answer No while the former will still answer Yes. Similarly, if a couple of type *NYY* are asked questions 1 and 2 respectively, they will answer No and Yes respectively.)

Although he didn't consider the (analogue of the) case where the dials have different settings, Einstein concluded from a similar situation (the Einstein, Podolsky, Rosen experiment—EPR for short) that there must be "hidden variables" (the *RRG, GRG*, etc.) that determine whether the detectors will flash red or green, and that will explain why both detectors *always* flash the same color when their settings are the same. Note that in the case of the married couples this conclusion certainly seems reasonable enough.

So far, this whole discussion may seem a bit out of place and not worth pursuing (or even beginning)—something like a long shaggy-dog story without a punch line. Well, the punch line is coming. I quote from N. David Mermin's article "Quantum Mysteries for Everyone," where he writes: "The apparent inevitability of this explanation (above) for the perfect correlations when the dial settings are the same forms the basis for the conundrum posed by the device. For the explanation is quite incompatible with what happens when the dial settings are different" (1981).

Consider a particle of type *GRG*. Out of the six possible dial settings that differ (12, 21, 13, 31, 23, 32), only 13 and 31 will result in the same color flashing (green in this case). Thus for this type of particle, *GRG*, the same color should flash one-third of the time. Similarly, for particles of types *GRR*, *RGR*, *RRG*, *RGG*, and *GGR* the same color should flash on the detectors at *A* and *B* one-third of the time. Particles of type *RRR* or *GGG* should always flash the same color when the dial settings differ. Thus we can conclude that the only reasonable account of why the colors are always the same when the dial settings are the same implies that when the dial settings are different, the colors flashed by the detectors should be the same (at least) one-third of the time—more if there are *RRR* or *GGG* particles. But empirically this just doesn't happen: when the settings differ, the colors flashed by the detectors are the same only one-fourth of the time.

(This contrasts with what happens in the case of the married couples. Consider one such couple, say of type *YNY*. Out of the six possible question settings that differ (12, 21, 13, 31, 23, 32), only 13 and 31 will result in the same answer being given (Yes for this couple). Thus for a couple of type *YNY*, the same answer should be given one-third of the time. Similarly, for couples of types *YNN, NYN, NNY, NYY*, and *YYN* the same answer should be given in Albuquerque and Buffalo one-third of the time. Since there are also some couples of type *YYY* or *NNN*, the same answer should actually be given more than one-third of the time.)

Weirdly enough, the fraction one-fourth is what quantum mechanics predicts, contrary to the above "common-sense" estimate of at least one-third developed in a slightly different context by the physicist J. S. Bell in 1964. (In the case of the married couples, given the quite unlikely assumption that couple members always do give the same answer when asked the same question, the estimate that at least one-third of the answers will be the same when the questions differ is, as we've seen, correct.)

Realism is the commonsense philosophical view that physical objects exist independently of being perceived. It was only considered to be a philosophy of science because a case against it had been made by idealists (e.g., Berkeley in the eighteenth century), who maintained that physical objects are mind-dependent, and their being consists in being perceived. Most red-blooded scientists and philosophers are still realists, but the quantum effects exhibited in microphysical (and macrophysical) phenomena put a good deal of new stress on this position.

There is no intelligible model or mechanism that explains these phenomena, there are just certain rules for calculating and predicting probabilities (as above, $\frac{1}{4}$ instead of $\geq \frac{1}{3}$). As David Mermin writes, "The device behaves as it behaves. . . . It is not the Copenhagen interpretation of quantum mechanics that is strange, but the world itself." Physicists either become slightly mystical, postulating some quality of "wholeness" to nature (in particular to the particle pairs being measured, or to the unconnected parts A, B, and C of the device itself), or they become hardheadedly positivistic, taking refuge in their rules and formulas and abjuring intelligibility or explanatory models altogether.

Some nonphysicists (and even a few physicists) have interpreted the above experiment as indicating the existence of telepathy or instantaneous communication. The standard so-called Copenhagen interpretation of quantum mechanics rules this out, but the experimental results can nevertheless be interpreted in this way. (Even if they are, however, the "telepathy" or "instantaneous communication" involved is not of a sort that should be very comforting to these people; absolutely no message can be sent via it. The original Einstein–Podolsky–Rosen thought experiment is a simpler version of the one I've sketched here. Thus, to put it acronymically, EPR does not imply ESP.)

I wrote in the first chapter that if one understands the relevant philosophical point, one gets the joke. Not fully understanding quantum mechanics, we can't laugh quite yet (at least not heartily) at the above or other quantum mechanical puzzles. Still, I think someday we will—though, contrary to Einstein, I think that God or nature or the Great Tortoise really does throw dice.

A jest unseen, inscrutable, invisible, as a nose on
a man's face.   —*William Shakespeare*

We dance round in a ring and suppose,
But the Secret sits in the middle and knows.
                                                 —*Robert Frost*

"In just 38 months, you can earn BIG PROFITS as a fully trained QUANTUM MECHANIC. Learn secrets of QUANTUM MECHANICS in your own home, in your spare time, without quitting your present job! The nation is crying for fully trained QUANTUM MECHANICS.

"You get professional equipment to learn with. You will receive a professional cyclotron, actual atoms, and a year's supply of Preparation A for your atomic piles. Only $675. Cheap when you consider how proud you'll be to hear your son say "My Daddy's a QUANTUM MECHANIC!"

—*anonymous, magazine parody*

On Assumptions

Risking cerebral whiplash, let me momentarily flash back 2500 years to Zeno of Elea. Zeno wrote that Achilles could never catch the turtle, since by the time Achilles had reached the place P_0 from which the turtle had started, the turtle would have advanced to P_1. When Achilles reached P_1, the turtle would have advanced to P_2; by the time Achilles had reached P_2, the turtle would have advanced to P_3; and so on. Achilles would have to cover an infinite number of such intervals before he could catch and surpass the turtle. We know now that the sum of an infinite number of (interval) lengths can be finite (e.g., $\frac{1}{2} + \frac{1}{4} + \frac{1}{8} + \frac{1}{16} + \frac{1}{32} + \ldots = 1$). Along with some theorems on infinite sets due to Georg Cantor, this knowledge resolves Zeno's paradox.

It's a reasonable conjecture that some comparable sort of conceptual advance will be necessary if we are to understand (and not just predict) quantum phenomena. The presuppositions of classical physics that will probably have to be modified concern the notion of "thinghood." What is it to be a thing? How is it that things persist through time? Are things distinct, or are they fuzzy? Are they independent, or somehow interconnected? It may very well turn out that we do not know very much about things.

The American philosopher Hilary Putnam has written that just as the development of non-Euclidean geometry made possible the easy expression of Einstein's relativity theory, so might the development of a quantum nonclassical logic make easier and more natural the expression of quantum theoretical insights about "things." The law of the excluded middle, for example, might be modified (or it might not).

Imagine a list, in some order, of all the infinitely many numbers between 0 and 10: 7.1, $2^{33}/_{49}$, π, $9^{112}/_{219}$, $5\sqrt{2}$, $\sqrt[3]{19}$, 2.86312, . . . , e^2, $5^{11}/_{103}$, ... Around the first number, place an interval of length $^1/_2$; around the second number, an interval of length $^1/_4$; around the third number, an interval of $^1/_8$; around the fourth, an interval of $^1/_{16}$, and so on. The sum of these intervals is $^1/_2 + {}^1/_4 + {}^1/_8 + {}^1/_{16} + {}^1/_{32} + . . . = 1$, yet to most people it would seem that these intervals cover all the points on the number line between 0 and 10. What assumption are they (or you) making that Georg Cantor didn't make?

George is running up the aisle of a train traveling at 20 mph. If he's running 10 mph with respect to the train, he's moving at 30 mph with respect to the ground. It's natural to assume that velocities always add in this way—but the theory of special relativity states that this is not the case for very large velocities. Likewise, it's natural to assume that of any two space-time events, one must necessarily precede the other for all observers—but again, it isn't so.

Martha and a big dog are standing at a bus stop. Waldo approaches them and asks if her dog bites. She assures him that her dog is very friendly and doesn't bite, whereupon Waldo pets the dog. The dog bites his arms and legs and thoroughly mauls Waldo, who screams at Martha, "I though you said your dog doesn't bite!" Martha responds quite innocently, "Oh, that's not my dog."

Waldo asked his doctor how to improve his relationship with his wife. The doctor advised him to take a ten-mile walk each night so he wouldn't be so irritable, and to call him in a month. When Waldo called the next month, the doctor asked him how things were with his wife. "Fine, I'm very relaxed, but I'm three hundred miles from home!"

The moral is obvious. Suppositions, assumptions, pre-suppositions, whatever you want to call them, are necessary in order to do science or to "do" life, but they can be misleading and even dangerous when made unthinkingly. Unfortunately (or fortunately), they usually *must* be made unthinkingly.

Once entrenched, theories can be difficult to displace. Just as Ptolemy added epicycle to epicycle to save his theory of planetary orbit, people tend to embellish and embroider any barely serviceable theory and often even prefer such baroque theories to simpler ones.

Dr. Paul Watzlawick (1977) relates a story about a relevant piece of research due to Professor A. Bavelas: Two subjects A and B are asked to try by trial and error to recognize the difference between healthy and sick cells. They can only respond "Healthy" or "Sick" to the slides that are shown them. They are told that a light will signal when they have answered correctly. In fact, however, only A's correct responses are always greeted by the light signal. Unknown to B, who is seated across the room (and who sees the same slides in the same order), B's responses are greeted by the light signal only when A responds correctly. Whether B responds "Healthy" or "Sick" has no effect on the reinforcing light signal that he sees.

Afterward, when asked to explain their "theory of healthy cells," A's theory is simple, concrete, and straightforward. B's ideas, on the other hand, are complex, convoluted, and elaborate. Most surprisingly, A is impressed by the "brilliance" of B's theory and in a subsequent trial does considerably worse than in his first trial, presumably having been influenced by B's abstruse (Ptolemaic) ideas.

Of course, no formal research is needed to note that, everything being equal, people are often more impressed by mumbo-jumbo that they don't understand than by simple observations and deductions that they do understand. They prefer hairy hypotheses to shaving with Occam's razor.

The mathematician Howard Eves (1958) recounts the tale of a man who loved to walk. His problem was that (like Waldo earlier) he often found himself far from home at nightfall. He thus decided to buy a house on the side of a big hill and walked around the hill once each day, starting in the morning with the rising sun behind him and ending back at his house at dusk with the setting sun ahead of him. After a few years of this he discovered to his horror that his uphill leg had shortened considerably. The man then decided to walk in the other direction for a few years until he evened the length of his legs. When neighbors doubted his story, he always responded by pulling up his trousers and saying, "Look, aren't my two legs the same length?"

Needless to say, therapists ("She had such and such a depressive disorder and now she's cured"), economic forecasters, political pundits ("There was a tremendous backlash, but now attitudes toward such and such have become stable"), and the proverbial man in the street (as well as the proverbial man on the hill) are all sometimes prone to support their diagnoses and "cures" with similar arguments.

Assumptions structure one's thinking, and hence at times obstruct the seeing of certain "obvious" phenomena and ensure the seeing of certain "nonexistent" phenomena. (The quotes indicate that what is obvious or nonexistent depends to some extent on one's theoretical assumptions.)

For example, chemists before Lavoisier did not observe many of the phenomena associated with rust and oxidation, since their theoretical assumptions concerning phlogiston made it unlikely that these phenomena would be noticed. Biologists before Harvey, though they could not find any, were convinced there were holes between the left and right halves of the heart, since their theories pointed to such a conclusion.

Scientists often behave like Van Dumholtz does in the following story, though the esoteric nature of their concerns helps to keep this relatively secret:

Van Dumholtz has two large jars before him, one containing many fleas, the other empty. He gently removes a flea from the flea jar, places it on the table before the empty jar, steps back, and commands "Jump," whereupon the flea jumps into the empty jar. Methodically he gently removes each flea, places it on the table, says "Jump," and the flea jumps into the originally empty jar.

When he has transferred all the fleas in this way, he removes one from the now full jar, carefully pulls off its back legs, and places it on the table before the original jar. He commands "Jump," but the flea does not move. He takes another flea from the jar, carefully pulls off its back legs, and places it on the table. Again he commands "Jump," but the flea does not move. Methodically he goes through this same procedure with the remaining fleas, and gets the same results.

Van Dumholtz beamingly records in his notebook: "A flea when its back legs are pulled off, cannot hear."

Finally, teleological explanations make reference to the end state or purpose of some phenomenon in order to explain that phenomenon. Such explanations, for example, were (and still are) used to counter Darwin's theory of evolution. Voltairean parodies of them—rabbits have white tails so as to be more easily hunted, noses have bridges to ensure the comfortable placement of eyeglasses, etc.—are well known. Much less known is the argument advanced by Hy Marx (Groucho's great-uncle and a famous teleological physiologist) to account for the foul odor associated with flatulence: the smell, Hy explained, was for the benefit of the deaf.

In contrast, there are also legitimate uses of teleological explanation, especially if such explanations ("the thermostat is trying to keep the house at a steady temperature," e.g.) can be reformulated in nonpurposive terms or, in more complex cases, in terms of the stability of systems with interacting parts. Kant, in fact, wrote that the ability to recognize purposiveness in nature, or teleological judgment as he called it, is intimately connected with "common sense."

Another legitimate variant of teleological explanation will be discussed in the next chapter.

chapter four

PEOPLE

Context, Complexity, and Artificial Intelligence

READ: Out of Sight, Out of Mind
PRINT: Blind Idiot

Researchers in the field of artificial intelligence have been concerned with the construction of programs to translate articles from one natural language to another. Earlier workers vastly underestimated the complexity of the task, however. Something of the flavor of the problems encountered is given by the following apocryphal story: An early Russian-English, English-Russian translating program took "The spirit is willing, but the flesh is weak," translated it into Russian, and then retranslated this Russian translation back into English. The result was "The vodka is agreeable, but the meat is too tender."

Whenever there is a metaphoric component to a passage, as above, or a dependence on context or background knowledge, similar problems arise. On the other hand, if the task if formal, if there are hard-and-fast rules to follow, as in the game of chess, the performance is, of course, much more impressive. In fact it is incomparably easier for a computer to determine trajectories for space vehicles, say, than it is for it to carry on an ordinary open-ended conversation (via video screens) with a human interlocutor. The latter feat is not even close to being accomplished. If

and when a computer is programmed with enough structured factual knowledge, memory, and self-modifiability to
carry on such a conversation in a manner indistinguishable
from that of another person, it will have passed (Alan)
Turing's test for machine intelligence.

To pass the Turing test, not only must an awesome
amount of informal and mundane knowledge be formalized
(mustard is not put on bananas or in one's shoes, cats do
not grow on trees, raincoats are not made out of rain) but
differences in significance due to context must also somehow be provided for. How, to cite a common example, is a
computer to evaluate a remark about a man touching his
head without knowing the context in which it occurs? It
could mean indefinitely many things, depending on indefinitely many ever-changing human contexts. Even if
enough knowledge of the most likely contexts is programmed in, it is commonly the case in conversation that
the context of a term is supplied only within the bit of dialogue in which it appears. A specific example is provided by
the following story, but this openness and sensitivity to context, this being-thereness, is pervasive.

A young man on vacation calls home and speaks to his
brother. "How's Oscar the cat?"

"The cat's dead, died this morning."

"That's terrible. You know how attached I was to him.
Couldn't you have broken the news more gently?"

"How?"

"You could've said that he's on the roof. Then the next

time I called you could have said that you haven't been able to get him down, and gradually like this you could've broken the news."

"Okay, I see. Sorry."

"Anyway, how's Mom?"

"She's on the roof."

The dog moves his rook to KB4 with his paw. George moves his queen to QB6 and announces, "Checkmate. That was a stupid move, dog. Besides having bad breath, you're really dumb. I've beaten you five out of seven games."

There is an interesting tendency that workers in artificial intelligence (the least critical, most gung-ho of whom are sometimes castigated as members of the "artificial intelligentsia" by disgruntled art historians) often complain about: Any activity or task—such as translating novel passages from one natural language to another—that is resistant to performance by a computer is defined as requiring "real" intelligence. Conversely, anytime a program is written that accomplishes some task or other, such as the many chess-playing programs, many people dismiss the task as not requiring real intelligence. This tendency to dismiss as not really constitutive of intelligence those activities formalized enough to be captured by a program should not, I think, be considered an after-the-fact redefinition of intelligence, but rather a reevaluation of the various kinds of intelligence.

Performing many different tasks, using many different procedures, determining when one rather than another is appropriate, discovering the limits of a method's applicability, modifying a system to adapt to different circumstances—these activities are all of a different order of difficulty and require a more flexible sort of integrative intelligence than does working within a well-defined formal system such as chess or celestial mechanics. Necessary in order for this "integrative intelligence" to function in putting together the disparate and incongruous details of an informal situation and making them coherent is a personality of sorts—wants, interests, a sense of self.

It is tempting to speculate that the high-status jobs of the future will be those that place a premium on these self-mediated integrative activities, rather than on any particular formalized skill. Generally, the more interesting and important a job, the less well-defined it is. Housewives and househusbands, handymen, humorists, philosophers, good conversationalists, friends, and lovers—all may be valued even more highly in the future than they are today.

Marvin Minsky, an eminent computer scientist, has written, "When intelligent machines are constructed, we should not be surprised to find them as confused and as stubborn as men on their convictions about mind-matter, consciousness, free will, and the like." This seems to make some sense, though I would substitute an "if" for the "when" above.

In addition to philosophical concerns, these intelligent "machines" will probably have a sense of humor. In fact, a variation of the Turing test for machine intelligence would be to construct a program that recognizes jokes. All the intellectual integrative skills mentioned earlier would be required, along with an appreciation of emotional nuance. This combination of skills is not so common, come to think of it, even among humans.

A very old married couple in their nineties visit a divorce lawyer. He asks, "Why now? You're both in your nineties, you've been married for more than seventy years, why get divorced now?"

They explain, "We wanted to wait until the children were dead."

If you chuckled, you probably don't have silicon in your brain (steel in your heart maybe, but not silicon in your brain).

It's conceivable that with the advance of artificial intelligence, ethnic jokes will be replaced by robot jokes:

Two robot bear hunters were driving along when they came upon a sign that said "Bear Left," whereupon they returned home.

The robot pharmacist quit his job. He couldn't fit the little prescription bottles into the typewriter.

An important distinction in computer science is that between the hardware and the software of the computer. Although the distinction is not always clear-cut, "hardware" refers to the physical aspects of the computer (tapes, disks, transistors, chips, etc.), whereas "software" refers to the programs that run on the computer. The program determines what the computer does, what the sequence of logical or programmatic states must be. Corresponding to these logical or programmatic states of the computer are the physical states of its hardware.

Hilary Putnam has noted that the logical and linguistic questions and issues that arise concerning this software-hardware distinction are similar in some important respects to those arising in the traditional mind-body problem of Descartes. What is the relationship between mind and brain (body)? How does one affect the other? Are the mental and physical incommensurable, or are they different aspects of the same phenomenon? These problems, Putnam claims, have in some respects solutions (or dissolutions) identical to those of the following analogous problems: What is the relationship between program and hardware? How does one affect the other? Are programmatic and hardware properties incommensurable, or different aspects of the same phenomenon?

Compare:

(1) I want George to cry at this point in the play, so while he's backstage either have him think of something very sad or, if he can't, drip onion juice into his eyes.

(2) I want that strange helical pattern to appear on the monitor at this point in the presentation, so either program its appearance or, if you can't, rub a magnet over the interface cable like this.

The topic of the next section (intensional explanations) sheds a little light on some related questions.

"And the problem arises: what is left over if I subtract the fact that
my arm goes up from the fact that I raise my arm?"

—*Ludwig Wittgenstein*

MYRTLE: Why do you think that man over there just touched
his head?

GEORGE: He's the third-base coach and he's giving the bunt
signal to the batter.

MARTHA: It's a windy day and he's making sure his hairpiece
is snug.

WALDO: A complex set of neuron firings and muscle con-
tractions brought about by an even more complex set of
chemical and physical phenomena caused the upper
right appendage to move at such and such an angle and
rate of speed to the lateral part of the uppermost central
extremity.

MYRTLE: Huh?

The explanations of George and Martha differ from
Waldo's in a crucial way: they explain by giving a reason for
the behavior in question, rather than by citing causal laws.
By giving a rationale for the behavior, George and Martha

make it reasonable in light of certain socially accepted rules and norms, and the beliefs and intentions of the agent. Explanations of this type, which presuppose the rationality of the agents involved, are called intensional explanations. Waldo's explanation, on the other hand, is a causal one: If these general covering laws are valid, and if these conditions obtain, then that will result.

Note that there is no conflict between the two types of explanation. Both types can be invoked to explain the same bit of behavior (Princess Diana's becoming pregnant, the Watergate tapes being erased), though one or the other may be more appropriate in any given context.

It's interesting that Freud, who started out as a "hard" scientist, attempted in his "Scientific Project" to reconcile causal (neurophysiological) explanation and intensional (psychoanalytic in this case) explanation. He wrote that he wanted "to furnish a psychology that shall be a natural science: that is, to represent psychical processes as quantitatively determinate states of specifiable material particles, thus making those processes perspicuous and free from contradiction" (1966). He failed, of course, since his neurological knowledge was limited, his psychoanalytic theories were flawed (to be kind), and the connection, even between two ideal approaches, was and is almost unimaginably complex.

Even in purely physical contexts, intensional explanations may sometimes be the only manageable ones. Playing chess with a computer, for example, requires that one adopt what Daniel Dennett (1978) calls an "intentional stance" toward it rather than a physical one. One predicts (and explains) the computer's moves by asking oneself what is the most rational move given the goal of the program (winning), the constraints on it (the rules of chess), its store of information (perfect memory of all past moves), and its "personality" as so far revealed (tendency to castle, move its queen, and so on). One does not try to predict or explain the computer's moves by examining the physical state of its circuits, chips, and disks, because to do so is again much, much too difficult, and prohibitively time-consuming.

The notion of an action is useful in clarifying the relation between intensional and causal explanations. An action is a bit of behavior for which the agent's reasons are the cause. The erratic movement of an epileptic is not an action, nor is the backward fall of a man sitting on a rickety chair that collapses suddenly; a causal explanation is most appropriate in these cases. On the other hand, jumping around energetically to alert an airplane overhead and a backward double somersault, though they resemble the previous behaviors, are both actions. It should be clear also from the example of a man touching his head that a bit of behavior can be seen as indefinitely many different actions depending on the person, his background, the immediate context of the behavior, and the general culture. The man might have touched his head in the belief that he would thus appear relaxed and unconcerned while his neighbors discussed a local burglary, a burglary the man himself committed. Causal explanations do not have the variety and dependence on context that intensional explanations do.

Though the reasons of an action are its cause, it is important to remember that they cannot be determined except in the context of an intensional explanation. One first determines what action a person is performing, why it is rational, and then identifies that person's reasons as the cause of the action.

My son often fights with my daughter, and sometimes in defense he explains, "I didn't punch her. My arm was just moving and her face was in the way."

The French writer Henri Bergson attributed laughter to the "mechanical encrusted on something living" (1911). By this rather celebrated phrase he meant that a person who becomes rigid, machinelike, and repetitive becomes laughable, since the essence of humanity is its (relative) flexibility and spirit. "Any incident is comic that calls our attention to the physical in a person when it is the moral side that is concerned. . . . We laugh everytime a person gives the impression of being a thing" (1911). Slipping, falling, digesting, twitching—none of these are actions, none admit of intentional explanations, and all, in a suitable context, are funny.

Hamlet hiccoughs, the president burps, or the mob boss slips on a banana peel, and we laugh. On a somewhat higher level, philosophers (which most people are, to some extent) laugh when someone makes a category mistake of some sort. Philosophy is devoted to the never-ending task of seeing beyond the "mechanical encrustations" on our understanding.

GEORGE: This talk of intentional explanations is sloppy. Why don't we use only causal explanations?

MARTHA: You're right. Let's just *decide* right now to do that. We both *want* to clarify, and causal explanations seem clearer and more precise.

The joke, such as it is, is that Martha is offering an intentional explanation of why she and George are planning to use only causal explanations. Intentional notions are built into the fabric of our communication. The American philosopher H. P. Grice has even analyzed "S's meaning something by x" as "S's intending the utterance of x to produce some effect in a hearer by means of the hearer's recognition of S's intention to produce that effect" (1957). For practical purposes, causal explanation correlates of intentional explanations are quite useless.

An important way in which the logic of intentional explanations differs from that of causal ones is in the failure of extensionality. That is, the substitution of an expression referring to a person or thing for another expression referring to that same person or thing can change the truth-value of intentional statements and the cogency of intentional explanations. This is not the case for causal statements or explanations. For example, consider again George's explanation of why the man was touching his head. (He's the third-base coach and he's giving the bunt signal to the batter.) Suppose that the third-base coach is Henry Malone's cousin and the batter is the only Greek on the field. Upon substitution of these expressions, the "explanation" of the man's touching his head becomes "He's Henry Malone's cousin and he's giving the bunt signal to the Greek." One could even replace "giving the bunt signal to" with "stroking his temple for," an expression referring to the same movement, and lose all explanatory power. The substitution of an expression having the same referent for an expression in Waldo's causal explanation—say, by using a different coordinate system, or a different description of "upper right appendage"—does not affect the truth-value of any statement, or the explanatory power of the explanation. Thus in intentional, but not in causal, explanations, how a person or thing is referred to or described is important. Even though Oedipus wanted to marry Jocasta, Jocasta was his mother, and presumably Oedipus did not want to marry his mother.

A confusion or conflation of subject and object always results in undecidable, open questions. Recall the brilliant computer whose yes-no device turned the attached light-bulb off and on. The subject-object conflation resulted in the undecidability by the machine of certain questions involving the lightbulb.

More commonly one forms a model of a situation, and if one is a part of that situation, one objectifies that part of oneself so involved. The account of the situation is then necessarily incomplete, however, since a part of the subject-observer is always doing the observing and that part is not being self-observed. This logical problem represents a "problem," of course, only when one wants a "complete" account or explanation of something, when one becomes too greedy a scientist. There's no problem if one is simply dancing or fighting, making love or practicing Zen, picking grapes or picking one's nose.

Intentional explanations in general involve such sub-ject-object blurs, since they require of one enough empathy to understand the rules, values, and beliefs of another person whose responses and actions are in turn thereby affect-ed. Compare Grice's account of communication above. Except in the case of quantum phenomena (Heisenberg uncertainty principle), causal explanations generally don't have this property. A rock is not affected by any calculations or explanations of its trajectory, for example.

GEORGE: Hi, Martha.

MARTHA: What's the matter, George? Are you mad at me?

GEORGE: No, of course not.

MARTHA: Yes you are. Why are you mad?

GEORGE: I'm not, I told you.

MARTHA: You are. I can tell by the tone of your voice.

GEORGE: Martha, I am trying not to be angry with you.

MARTHA: See, you're seething with hostility toward me. Why? What did I ever do to deserve such anger?

George stalks away, slamming the door behind him.

The *National Inquirer*, a strange yet intriguing gossip-filled tabloid of news and pseudo-news, often carries stories about why show-business celebrities X and Y are in love. The stories, I would guess, often play a role in the subsequent appearance of stories about why celebrities X and Y are now involved with Z and W, respectively. Even "neutral, objective" observation of people often significantly affects their behavior; consider surveys of the frequency of various sexual practices, for example. However, people sometimes overestimate the effect that their doings will have on others. Such may be the case with the confused statistician in the following story:

Howard Eves (1958) tells about a statistician who traveled widely giving lectures. He was, however, apprehensive about flying, especially since there had been some recent bomb scares aboard airplanes. He calculated the probability of a bomb being aboard a plane and was reassured that it was reasonably small. Then he calculated the probability of there being two bombs aboard a plane and found it to be absolutely infinitesimal. Hence he always traveled with a bomb in his suitcase.

Intentional explanations are probabilistic for several reasons. First is the subject-object blur just discussed. Explaining, even observing, often changes what is explained or observed, certainly with respect to oneself and those close to oneself. The second reason is the nature and level of the explanation. Providing a plausible, broad-scale rationale for an action is not the same as providing a set of sufficient causal conditions for it. A (or many) rationale(s) for a potential action is (are) never sufficient to ensure that the action will take place. No matter how many cogent reasons there are to do something, a person (or other intelligence) may still decide not to do it.

Microphysical indeterminism may, as I've already noted, be a third source for the probabilistic nature of intentional explanations. It is conceivable that this quantum indeterminism is somehow filtered through the structured complexity of the mind-brain, resulting in the possibility of metaphysically free action. (Bertrand Russell once speculated, "It might be that without infringing the laws of physics, intelligence could make improbable things happen, as Maxwell's demon would have defeated the second law of thermodynamics by opening the trap door to fast-moving particles and closing it to slow-moving particles.") Ideas (some unconscious) might be generated in part by an indeterministic process, the viable ones surviving the tests of reality and consciousness. The similarity to biological natural selection is especially appealing. In any case, this third source is not required, as even intentional explanations or predictions of the moves of fully determined chess-playing computers are still, practically speaking, only probabilistic.

Finally, it should not be assumed that the value of an explanation is measured by how probable its conclusion is, as George shows Waldo:

WALDO: The value of an explanation is only as good as the probability of its conclusion.

GEORGE: What am I doing?

WALDO: It looks like you're rolling a pair of dice.

GEORGE: You're right. I just rolled a seven. Why?

WALDO: Chance. One-sixth of the time you'll roll a seven, that's all.

GEORGE: Now I've rolled a twelve. Why?

WALDO: Chance again. One-thirty-sixth of the time you'll roll a twelve.

GEORGE: The explanations seem to be the same even though the probabilities of their conclusions differ.

In intentional explanations as well as causal ones, a low probability for the conclusion does not necessarily mean a poor explanation. Our specific, personal genetic makeup, for example, is an improbable accident: a different sperm might have united with the same or a different egg and we wouldn't have been. Still, the explanation for our personal genetic makeup depends on the particular sperm and egg that did, however improbably, unite.

Arrow, Prisoners, and Compromise

> A young boy is quietly masturbating in his room when his mother walks in. "Don't do that, son, or you'll go blind."
> "Mom, couldn't I do it just until I need glasses?"

> Two dangers threaten the world—order and disorder.
>
> —*Paul Valéry*

Moderation has been recognized as a virtue since the time of the Greeks. The problem with the injunction "Be moderate," however, is that it's almost meaningless without some standard scale against which to measure oneself and one's actions. Even in elementary mathematics, "Be moderate" faces difficulties. If the side of a cube ranges between 0 and 2 inches, then, assuming we know nothing else of the distribution of cube sizes, the average cube might be said to have a side of 1 inch. Yet if instead we consider the volumes of these cubes, which must range between 0 and 8 cubic inches, we might say that the average cube has a volume of 4 cubic inches. Thus, using two different standards, the "average" cube has a side of 1 inch and a volume of 4 cubic inches.

Late in the eighteenth century Jeremy Bentham tried to develop a "hedonic calculus" that would measure the utilities and disutilities of an action. With the aid of such a tool we could "act so as to maximize the net amount of utility" (1948). How then does this wonderful calculus measure utilities and disutilities, or, more plainly, pleasure and pains? Unfortunately (or more likely, fortunately), it doesn't exist. Bentham and others since him have failed to find any reasonable way to add up the incommensurables of human life. (There is no trick in finding unreasonable or arbitrary ways, or even reasonable ways in narrow contexts.)

Though Bentham said little about how to compare the utility of disparate qualities—beauty vs. intelligence, wealth vs. health—he did suggest that for any one quality, "extent," "duration," and "intensity" were reasonable dimensions to consider. They are, of course; but problems remain. How does one measure even these dimensions of a quality? For what duration, for example, does a conversation with a friend afford one happiness—a half-hour, a week, a lifetime? How are the values of these dimensions related to the quality in question? What is the relation between the number of jokes, say, and the humorousness of a book? How do these dimensions interact with each other and with other more subtle dimensions of a quality? Can interpersonal comparisons of intensity be made? Does it make sense to say that George loves his son more than Martha loves her husband? In short, not only is there no way to compare apples with oranges, there is not even an all-purpose way to grade apples.

Still, there are *special-purpose* ways of grading apples—weight, sugar content, skin thickness, number of worms, and so on. Likewise, there are special-purpose ways of grading beauty (hair color, skin oiliness, size of ears, and so on), intelligence (special intuition, vocabulary, memory, and so on), health (weight, various blood counts, number of operations, and so on). As long as these special-purpose orderings and rankings are recognized as rough, circumstantial, and conditional, much of the crassness of such rankings can be avoided.

MARTHA: Is he smart?

MYRTLE: Yeah, his IQ is 160.

MARTHA: Is he rich?

MYRTLE: He makes $400,000 a year.

MARTHA: Is he good-looking?

MYRTLE: He makes Paul Newman look like Rodney Dangerfield.

MARTHA: Is he friendly, sexy?

MYRTLE: Everyone knows him. He's always at some party or other, girls draped all over him.

MARTHA: Do you like him?

MYRTLE: I can't stomach the sight of him. He's plastic, phony, and insensitive. Just hearing his voice makes my flesh crawl.

GEORGE: Sure is hot today, about 80° I'd say.

WALDO: Yeah, got down to 40° last night. It's twice as hot now.

On the subject of hot summer days, there was a midwestern state legislator who opposed the adoption of daylight savings time because the increased daylight would more quickly fade curtains and fabrics.

That balance and moderation, a sense of perspective, and the harmonious integration of disparate elements are valued is evidenced by the laughter occasioned by their absence. Exaggeration and distortion, in particular, often lead to humor. The following is an old story due to George Bernard Shaw. It seems more appropriate with Groucho, however:

GROUCHO (*to woman seated next to him at an elegant dinner party*): Would you sleep with me for ten million dollars?

WOMAN (*giggling*): Oh, Groucho, of course I would.

GROUCHO: How about doing it for fifteen dollars?

WOMAN (*indignant*): Why, what do you think I am?

GROUCHO: That's already been established. Now we're just haggling about the price.

Politics is the art of balance and compromise. Consider the fact that the two most important ideals—liberty and equality—are, in their purest form, incompatible: complete liberty results in inequality, while enforced equality results in loss of liberty. Liberty and equality must thus be balanced, refined, made conditional. The differences between the political right and the political left can be simplistically described as a difference between prepositions, between freedom *to* (speak, move about, buy and sell, etc.) and freedom *from* (hunger, joblessness, etc.). The following pair of stories point up a couple of these differences:

George and Waldo come upon a couple of apples, one large, the other small. George, being quicker, grabs the larger apple and gobbles it up, while Waldo just manages to get hold of the smaller one.

WALDO: "That's not polite, George. If I'd reached here first, I would have left the larger apple for you."

GEORGE: "Then what are you complaining about? You got what you wanted."

George, Martha, Waldo, and Myrtle come upon a loaf of bread. They decide to divide it evenly, and use the following procedure: George cuts off what he considers to be a quarter of the loaf. If Martha thinks the piece is a quarter of the loaf or less, she doesn't touch it; if she thinks it's bigger than a quarter of the loaf, she cuts off a sliver to make it exactly a quarter. Waldo then either leaves the piece alone or trims it further if he thinks it's still bigger than a quarter of the loaf. Finally, Myrtle has the same option: trim it if it's too big, or leave it alone if it's not. The last person to touch the slice keeps it. This finished, there are three people remaining who must divide the rest of the cake evenly. The same procedure is followed: the first person cuts off what he considers to be a third of the remaining loaf, and so on. In this way everyone is satisfied that he or she has received a quarter of the loaf.

The following story is due to Raymond Smullyan (1980):

GEORGE: Mmm, chocolate cake. I'm going to eat all of it.

MARTHA: I want dessert, too. We should split it 50-50.

GEORGE: I want all of it.

MARTHA: No, we have to split it evenly. Let's ask Myrtle to decide. She's always fair.

MYRTLE: You should compromise: three-fourths for George, and one-fourth for you, Martha.

Mort Sahl remarked that in the 1980 election many people did not vote for Ronald Reagan so much as they voted against Jimmy Carter. He continued, "If Reagan had been unopposed, he would've lost."

People have individual preferences out of which, since these preferences often differ, group preferences must be fashioned. This is obviously a difficult practical problem. It is a more difficult theoretical problem, in a certain sense even an impossible one.

Consider first a voting paradox due to the eighteenth-century French philosopher Condorcet. Three candidates—

George, Martha, and Waldo—are running for governor of Wisconowa. A third of the electorate prefers George to Martha to Waldo; another third of the electorate prefers Martha to Waldo to George; and the remaining third prefers Waldo to George to Martha. There is nothing especially unusual about this unless we consider what happens in two-person races given the above preferences. George can boast that two-thirds of the electorate prefers him to Martha. Waldo responds that two-thirds of the electorate prefers him to George. Finally, Martha counters by noting that two-thirds of the electorate prefers her to Waldo.

If the societal preferences in this example are determined by majority vote, we have an irrational societal ordering of preferences—that is, "society" prefers George over Martha, Martha over Waldo, and Waldo over George. Thus even if the preferences of all the individual voters are transitive (transitivity holds if, whenever a voter prefers x to y and y to z, he or she prefers x to z), the societal preferences determined by majority vote are not necessarily transitive (rational).

A general theorem can be proved showing that all "reasonable" voting systems (or equivalently, economic market systems) are subject to such irrationalities; but before discussing this, it should be pointed out that individuals are not immune to Condorcet's paradox.

The mathematician Paul Halmos has proposed the following variation of the paradox that applies to individuals: Imagine a woman trying to decide which of three cars to buy: Car G, Car M, or Car W. She, being a methodical sort, had three criteria (of equal weight) for making this decision: looks, affordability, and performance. Car G looked better than Car M, which looked better than Car W. On the other hand, Car M was more affordable than Car W, which in turn was more affordable than Car G. Finally, Car W performed better than Car G, which performed better than Car M. Since the woman placed equal weight on each of these criteria, she was in a quandary. She clearly preferred Car G to Car M (G outscored M on two criteria). She also preferred Car M to Car W (for the same reason). Yet she preferred Car W to Car G.

Though the same problem of nontransitivity holds for individuals, it seems somehow more tractable. In the case above you just have to induce the woman to declare one of the criteria more important than the others. This is easier than convincing one-third of an electorate to change its mind.

The economist Kenneth J. Arrow proved (1951) an interesting generalization of Condorcet's paradox. He showed that there is no way to derive societal or group preferences from individual preferences that can be guaranteed to satisfy the following four minimal conditions: The societal preferences (1) must be transitive (if society prefers x to y and y to z, then it must prefer x to z); (2) must satisfy the Pareto principle (if alternative x is preferred to alternative y by everyone in the society, then society must prefer x to y); (3) must satisfy the independence of irrelevant alternatives (the societal preference depends only on the orderings of the individuals with respect to alternatives *in* that environment); and (4) must not be susceptible to dictatorship (there is no individual whose preferences automatically determine all of society's preferences).

COMMUNIST: "Man's inhumanity to man," that is what capitalism is all about.

GEORGE: Yes, in communism it is the other way around.

Bumper sticker: GOD SAID IT, I BELIEVE IT, AND THAT SETTLE'S IT. (The apostrophe is most informative.)

There was once a scorpion who wanted to cross a river. Spotting a turtle on a rock, the scorpion asked if he would take him across the river. In return he would show the turtle where he could find some very succulent vegetation.

The turtle responded, "How can I be sure you won't sting my neck?"

"Don't be silly," the scorpion replied. "If I did that, we'd both drown."

The turtle was convinced, and they began swimming toward the vegetation the scorpion had told him about. When they had nearly reached the far side of the river, however, the scorpion stung him in the neck after all. Struggling to make it to the river bank, the turtle gasped,

"Why, why did you do this to me?"

Hopping off the dying turtle's back, the scorpion explained, "I thought you might dive under the water to drown me."

The Prisoners' Dilemma is another old puzzle with societal implications. Imagine two prisoners Waldo and George against whom there is little real evidence. Although these prisoners have committed a crime, they can expect little punishment (say one year in prison) if they both remain silent. If George confesses to their crime, however, and Waldo remains silent, George will be released while Waldo can expect a five-year sentence. Conversely, if Waldo talks and George remains silent, Waldo will be released while George can expect a five-year sentence. Thus each man has a choice of confessing or not. If they both confess, each gets a three-year term; if they're both silent, each gets a one-year term; if *A* confesses and *B* is silent, *A* gets off and *B* gets five years.

The punchline, so to speak, is that the optimum course of action, remaining silent, is in general not the course that George and Waldo are likely to follow: they're both likely to confess so as to avoid being a patsy for the other. This situation is, of course, not limited to prisoners. Spouses in a marriage, businessmen in a competitive market, and national governments in an arms race can all fall subject to such prisoners' dilemmas. Adam Smith's invisible hand ensuring that individual pursuits will necessarily ensure group well-being is at times quite arthritic.

Results like Arrow's paradox and the Prisoners' Dilemma, the well-known problems of quantification and measurement in the social sciences (6.28 on the pleasure scale, 2.89 on the pain scale, let's do it), and the inherent riskiness of all explanation, especially intentional explanation, should, but probably won't, make for cautious skepticism in predicting ourselves, society, and the future. Lewis Carroll's Cheshire Cat reminds us of the single most important, yet unpredictable, determinant of that future:

"Would you tell me please, which way I ought to go from here," asked Alice.
"That depends a good deal on where you want to get to," said the Cat.

AFTERWORD

> When I was young, I forgot to laugh. Later when I opened my
> eyes and saw reality I began to laugh and haven't stopped
> since.   —Soren Kierkegaard

Wittgenstein once remarked that he looked forward to the day when philosophy was no longer a subject in its own right but rather infused all other subjects. Philosophy is (or should be), in this view, an adverb: one does linguistics philosophically, one studies science philosophically, one investigates political issues philosophically. Humor or play has something of the same character. It's awkward for humor itself to be the focus of an activity. The announcement "We will now tell jokes and be humorous" sounds distinctly totalitarian. Humor too is adverbial and qualifies one's approach to other activities: one answers questions humorously, analyzes a situation humorously, writes or speaks humorously.

Of course, "quickly," "painfully," and "odoriferously" are also adverbs, but I hope I've managed to indicate that "philosophically" and "humorously," at least in their best manifestations, share more than adverbial status. Both require a free intelligence in a relatively open society, and both evince a keen concern for language and its (mis)interpretation, as well as a skeptical tendency toward debunking. The incongruity that lies at the heart of most jokes is analogous to the conundrum that lies at the heart of most philosophical problems. Likewise, the aggressive tone present in many jokes and the social control that the jokes tend to foster are analogous to the argumentative nature of many philosophical papers and the intellectual dominance that the papers are meant to establish. It should be noted, though, that this aggressive tone and argumentative nature are clearly circumscribed and presuppose an independent intelligence in others.

Finally, both humor and philosophy are quintessentially human, requiring as they do the characteristically human ability to transcend one's self and one's situation. The discrepancy between our hopes or pretensions and reality is, try though we sometimes do, impossible not to see. Two responses to the starkness of this discrepancy are through philosophy and humor. I think, therefore I laugh.

Our heroes—Ludwig Wittgenstein, Bertrand Russell, Lewis Carroll, and Groucho Marx—are discussing the question "What is philosophy?" Let us join them for the end of their discussion and ours.

WITTGENSTEIN: I repeat, the question is not well posed. There are a whole family of uses for the term "philosophy." Still, my primary aim has been to clarify, to show how to pass from a piece of disguised nonsense to something that is obvious nonsense. Misunderstandings, as I've always insisted, must be cured if we are to be free of them.

GROUCHO: Is that anything like curing hams?

RUSSELL: Asking the question indicates, I think, that you know quite well what Mr. Wittgenstein is talking about.

GROUCHO: Easy on the self-reference shtick, Bertie. We're

not in the elevator anymore. I understand Ludwig's point, but most of this other philosophical stuff you guys dream up seems so nit-picking and technical. What about the big questions: the meaning of life, the death of God, the residuals on my television reruns?

RUSSELL: Better some real progress on the meaning of confirmation and probability, on the nature of logic and scientific law, on reductionism, artificial intelligence, and intentional explanation, for example, than a lot of empty blather on the so-called big questions. The big questions, at least the ones that make sense, will always be there. They're sometimes clarified by the answers to the smaller questions, sometimes not. When they're not, though, listening to woolly-headed pontificators expound on them doesn't help either. A courageous acknowledgment of ignorance is much preferable.

GROUCHO: Calm down, Bertie. Without those woolly-headed pontificators we might both be unemployed or, worse yet, lawyers. Maybe what I'm trying to get at—or be cured of, as Ludwig might say—is what difference does it all make? What difference will the answers to these smaller questions, or anything else for that matter, make in fifty thousand years? Even my reruns won't be on then, and there'll be no more copies of *Principia Mathematica*, Ludwig's *Philosophical Investigations*, or even *Alice in Wonderland*.

LEWIS CARROLL (OVERCOMING HIS SHYNESS, STAMMERS): Maybe n-n-nothing we do now will make a difference in fifty thousand years, but *if* that is so, then it would seem that nothing that will be the case in fifty thousand years makes a difference *now*, either. In particular, it doesn't make a difference now that in fifty thousand years what we do now won't make a difference.

GROUCHO: You can bet your walrus that I'm not going to tangle with you about time. The time has passed, in fact, to talk of many things. Enough. If you fellas will excuse me, I'm going to be leaving in a minute. If you persist with this talk, though, I might leave in a huff, or maybe even in a minute and a huff. In any case, since you haven't asked, I'm going over to make love with our old friend Martha. Fortunately her husband George is out looking for grue emeralds with Waldo.

Groucho slouches off, leaving the three philosophers wondering.

BIBLIOGRAPHY

Anobile, Robert J., ed. 1971. *Why a Duck*. New York: Darien House.

Arrow, K. J. 1951. *Social Choice and Individual Values*. New York: Wiley.

Barker, Stephen. 1964. *Philosophy of Mathematics*. Englewood Cliffs, N.J.: Prentice-Hall.

Bateson, Gregory. 1958. The Message "This Is Play." In B. Schaffner, ed., *Group Processes: Transactions of the Second Conference*. New York: Josiah Macy Jr. Foundation.

Bell, J. S. 1964. *Physics*.

Bentham, Jeremy. 1948. *An Introduction to the Principles of Morals and Legislation*. New York: Hafner.

Bergson, Henri. 1911. *Laughter: An Essay on the Meaning of the Comic*. New York: Macmillan.

Bohm, D. 1951. *Quantum Theory*. Englewood Cliffs, N.J.: Prentice-Hall.

Brody, Baruch. 1970. *Readings in the Philosophy of Science*. Englewood Cliffs, N.J.: Prentice-Hall.

Carroll, Lewis. 1946. *Alice's Adventures in Wonderland* and *Through the Looking Glass*. New York: Grosset and Dunlap.

Chaitin, Gregory. 1965. Randomness and Mathematical Proof. *Scientific American*, March.

Chaitin, Gregory. 1966. Complexity Theory. *Communications of the ACM*, August.

Davidson, Donald. 1963. Actions, Reasons, and Causes. *Journal of Philosophy* 60.

Dawkins, Richard. 1976. *The Selfish Gene*. New York: Oxford University Press.

DeLong, Howard. 1970. *A Profile of Mathematical Logic*. Reading, Mass.: Addison-Wesley.

Dennett, Daniel. 1978. *Brainstorms*. Vermont: Bradford Books.

Descartes, René. 1977. Meditations on First Philosophy. In *Classics of Western Philosophy*. Indianapolis: Hackett.

Dretske, Fred. 1971. Conclusive Reason. *Australasian Journal of Philosophy* 49.

Enderton, Herbert. 1972. *A Mathematical Introduction to Logic*. New York: Academic Press.

Eves, Howard. 1958. *Mathematical Circles Adieu*.

Farzan, Massud. 1973. *Another Way of Laughter*. New York: E. P. Dutton.

Frege, Gotlob. 1949. On Sense and Nominatum. In Herbert Feigl and Wilfrid Sellars, eds., *Readings in Philosophical Analysis*. New York: Appleton-Century-Crofts.

Freud, Sigmund. 1966. Project for a Scientific Psychology. In *The Standard Edition of the Complete Psychological Works of Sigmund Freud*. London: Hogarth Press and Institute for Psychoanalysis.

Fry, W. F. 1963. *Sweet Madness: A Study of Humor*. Palo Alto, Calif.: Pacific Press.

Gardner, Martin. 1973. Mathematical Games. *Scientific American*, July.

Gardner, Martin. 1981. *Gotcha*. San Francisco: Freeman Press.

Gettier, Edmund L. 1963. Is Justified True Belief Knowledge? *Analysis* 23.

Goodman, Nelson. 1965. *Fact, Fiction, and Forecast*. New York: Bobbs-Merrill.

Grice, H. P. 1957. Meaning. *Philosophical Review*.

Hempel, Carl. 1965. *Aspects of Scientific Explanation*. New York: Free Press.

Hofstadter, Douglas. 1982 and 1983. Metamagical Themas Column. *Scientific American*, January issues.

Hume, David. 1977. An Inquiry Concerning Human Understanding. In *Classics of Western Philosophy*. Indianapolis: Hackett.

Kant, Immanuel. 1977. Prolegomena to Any Future Metaphysics. In *Classics of Western Philosophy*. Indianapolis: Hackett.

Kripke, Saul. 1975. Outline of a Theory of Truth. *Journal of Philosophy*.

MacKay, D. M. 1964. Brain and Will. In *Body and Mind*. London: Allen and Unwin.

Malcolm, N. 1958. *Ludwig Wittgenstein: A Memoir*. London: Oxford University Press.

Margolis, Joseph. 1978. *An Introduction to Philosophical Inquiry*. New York: Knopf.

Mermin, N. David. 1981. Quantum Mysteries for Anyone. *Journal of Philosophy*.

Monk, Ray. 1983. *Ludwig Wittgenstein: Duty of Genius*. New York: Penguin.

Nozick, Robert. 1981. *Philosophical Explanations*. Cambridge, Mass.: Harvard University Press.

Nozick, Robert. Newcombe's Problem and Two Principles of Choice. In *Essays in Honor of Carl G. Hempel*. Dordrecht: Reidel.

Pagels, Heinz R. 1982. *The Cosmic Code*. New York: Simon and Schuster.

Pascal, Blaise. 1966. *Pensées*. London: Penguin Books.

Paulos, John A. 1980. *Mathematics and Humor*. Chicago: University of Chicago Press.

Pitcher, George. 1966. Wittgenstein, Nonsense, and Lewis Carroll. *Massachusetts Review*.

Poincaré, Henri. 1913. *The Foundations of Science*. New York: Science Press.

Popper, Karl. 1959. *The Logic of Scientific Discovery*. London: Hutchinson.

Popper, Karl. 1972. *Objective Knowledge*. Oxford: Oxford University Press.

Putnam, Hilary. 1975a. The Logic of Quantum Mechanics. In *Mathematics, Matter and Method*. Cambridge: Cambridge University Press.

Putnam, Hilary. 1975b. Minds and Machines. In *Mind, Language and Reality*. Cambridge: Cambridge University Press.

Quine, W. V. O. 1953. Two Dogmas of Empiricism. In *From a Logical Point of View*. Cambridge, Mass.: Harvard University Press.

Quine, W. V. O. 1960. *Word and Object*. Cambridge, Mass.: MIT Press.

Reichenbach, Hans. 1949. On the Justification of Induction. In Herbert Feigl and Wilfrid Sellars, eds., *Readings in Philosophical Analysis*. New York: Appleton-Century-Crofts.

Rosten, Leo. 1968. *The Joys of Yiddish*. New York: McGraw-Hill.

Russell, Bertrand. 1956. On Denoting. In R. C. Marsh, ed., *Logic and Knowledge*. London: Allen and Unwin.

Russell, Bertrand. 1924. *Introduction to Mathematical Philosophy*. New York: Macmillan.

Russell, Bertrand and A. N. Whitehead. 1910. *Principia Mathematica*. Cambridge: Cambridge University Press.

Salmon, Wesley. 1977. A Third Dogma of Empiricism. In *Basic Problems in Methodology and Linguistics*. Dordrecht: Reidel.

Skyrms, Brian. 1966. *Choice and Chance*. Belmont, Calif.: Dickenson.

Smullyan, Raymond. 1980. *This Book Needs No Title*. Englewood Cliffs, N.J.: Prentice-Hall.

Turing, Alan M. 1950. Computing Machinery and Intelligence. *Mind* 59.

Watzlawick, Paul. 1977. *Behavior and Paradox*.

Wittgenstein, Ludwig. 1953. *The Philosophical Investigations*. Oxford: Blackwell.

Wittgenstein, Ludwig. 1961. *Tractatus Logico-Philosophicus*. Translated by D. F. Pears and B. F. McGuinness. London: Routledge and Kegan Paul.

INDEX